M000304467

IT'S ~~NOT~~ EASY
BEING GREEN

IT'S ~~NOT~~ EASY BEING GREEN

HOW CONSCIOUS CONSUMERS AND ECOPRENEURS CAN SAVE THE WORLD

AVIVA PELTIN

NEW DEGREE PRESS

COPYRIGHT © 2019 AVIVA PELTIN

All rights reserved.

IT'S ~~NOT~~ EASY BEING GREEN

*How Conscious Consumers and Ecopreneurs
Can Save the World.*

ISBN   978-1-64137-309-8   *Paperback*

          978-1-64137-598-6   *Ebook*

MEDICAL DISCLAIMER

The material contained in this book is not intended to be, and cannot be taken as, a substitute for the advice and counsel of one's physician and/or other professional healthcare providers. The author is not a healthcare professional, and the reader should not base any medical decisions on any information in this book. For that reason, the author, publisher, people interviewed, and anyone associated with this book shall not be liable for any loss, injury, or damage allegedly arising from any information or suggestion in this book.

CONTENTS

"The most dangerous phrase in the language is, 'We've always done it that way.'"

—GRACE HOPPER

A CRUNCHY
UPBRINGING

———

Crunchy: Adjective. Used to describe persons who have adjusted or altered their lifestyle for environmental reasons .... *"I'm heading out to the crunchy store to pick up some fair trade chocolate and tea tree oil, do you need anything?"*

- URBAN DICTIONARY[1]

My childhood was not exactly conventional. We ate organic food, our juicer sat proudly atop the kitchen counter, words like quinoa and stevia were part of my vernacular, and a significant percentage of my wardrobe was tie-dyed. You

———

1   EBC. "Crunchy." Urban Dictionary. Urban Dictionary, June 14, 2007.

could say I was destined for crunchy greatness. My mom had turned to natural living when she had experienced health issues years before, so my father had been initiated into the crunchy lifestyle before I was even born.

I knew nothing else.

However, I wasn't totally on board with the crunchy life.

Now, I know Mom was just trying to keep me as healthy as possible, but at the time, natural and health-conscious living felt unfair. I just wanted to be like the cool kids. I would go to my friends' houses and drink Diet Coke, wondering why I couldn't do it at home. I would go to the drugstore and buy make-up with many ingredients that I now know the Food and Drug Administration (FDA) had not properly screened for safety. While some kids sneaked alcohol, I once bought and hid a conventional deodorant in my dresser drawer because effective natural deodorants did not exist.

That's normal, right?

I didn't understand my mom's reasoning behind our lifestyle choices, so they all felt unnecessary.

But then I got very sick toward the end of my adolescence. I was diagnosed with chronic Lyme disease and mold illness. My immune system was and is still struggling.

Ironically, I was the one to experience health issues, despite the fact that I was protected from so much. While I regret guzzling that Diet Coke and using that aluminum-laden deodorant, I was a kid and didn't know any better. Plus, health issues are cumulative—a complicated blend of factors and genetics that contribute to disease. However, this book is not about genetics or the nitty gritty biology of chronic conditions. I'm still trying to figure out how to regain my health, but I know the lifestyle changes that I have incorporated can do nothing but help, and if I hadn't altered my routines, I'd be feeling even worse.

I have spent hours upon hours researching treatments to try, toxins to avoid, products to use, and detox protocols to implement. I have learned so much, and if all my research can help you in any way, then I am delighted to write this book.

ABOUT THIS BOOK

This book is about how the products we apply to our skin, the food and drink we consume, the containers from which we drink, and even how we clothe our bodies, all add to our toxic burden or lessen it. This book is about how to be a smarter, healthier, and more sustainable consumer. This book is about questioning what medical professionals and institutions have propagated, and instead thinking for yourself.

This book also explores "ecopreneurship"—pioneering brands that are changing the world for the better. You will read exclusive interviews with CEOs and founders of successful companies that help to make a smarter, healthier, and more sustainable world. Many of them follow "triple bottom line" practices, which is an economic framework that focuses not only on profit but also on people and the planet.[2]

I hope these determined professionals inspire you as much as they have inspired me. This book is full of stories, insights, and lessons from these trailblazers:

- Jeffrey Hollender, co-founder of Seventh Generation, a company that produces cleaning, paper, and personal care products. He is also the current CEO and founder of The American Sustainability Business Council, an organization fighting for triple bottom line business practices.

- Barry Perzow, founder of Pharmaca Integrative Pharmacy, which blends conventional medicine with alternative modalities.

- Elise Graham Kennedy, founder of Honey and Vinegar, a non-toxic make-up brand.

- Pam Marcus, co-founder of Lifefactory, a glassware baby bottle and water bottle company inspired by Pam's experience working as a neonatal physical therapist.

---

2   Kenton, Will. "Triple Bottom Line (TBL)." Investopedia. Investopedia, May 3, 2019.

- Natalie Rechberg, founder of Daysy, a revolutionary fertility tracker that helps women plan or prevent pregnancy.

- Meika Hollender, founder of Sustain Natural, a company that makes safe and sustainable sexual health products.

- Antonia Saint Dunbar, co-founder of Thinx, a period-proof reusable underwear that many believe began the "Femtech" movement.

- Linda Balti, co-founder of Amour Vert, a sustainable clothing company.

- Mike Dally, founder of Earth Runners, a minimalist grounding shoe company.

- Cate Hardy, the CEO of Seattle-based PCC Community Markets, an organic and sustainable grocery store chain.

- Ryan Lewis, founder of EarthHero, a healthy and sustainable e-commerce brand.

- Rich Bergstrom, founder of Xtrema, a healthy cookware company.

- Lindsay McCormick, founder of Bite Toothpaste Bits, a zero-waste, all-natural alternative to conventional toothpaste.

- Tim and Kristi Zimmer, founders of Tummy Temple, a detoxification center in Seattle.

- Brian Richards, founder of SaunaSpace saunas, a non-toxic sauna company.

- Abigail Forsyth, co-founder of KeepCup, a reusable coffee cup company.

ENVIRONMENT AND HEALTH

Doctors take an average of seventeen years to incorporate new information and research into their practices,[3] so shouldn't we become informed and take charge of our own bodies before waiting for almost two decades for medical authorities to catch up? In 2019, researchers discovered that the common chemical, avobenzone, found in conventional sunscreens, penetrates the skin and enters the bloodstream.[4] This chemical is potentially carcinogenic and has been known to cause allergies and can age skin

---

3   Morris, Zoë Slote, Steven Wooding, and Jonathan Grant. "The Answer Is 17 Years, What Is the Question: Understanding Time Lags in Translational Research." Journal of the Royal Society of Medicine. Royal Society of Medicine Press, December 2011.

4   Matta, Murali K, Robbert Zusterzeel, Nageswara R Pilli, Vikram Patel, Donna A Volpe, Jeffry Florian, Luke Oh, et al. "Effect of Sunscreen Application Under Maximal Use Conditions on Plasma Concentration of Sunscreen Active Ingredients: A Randomized Clinical Trial." JAMA. American Medical Association, June 4, 2019.

faster.[5] Ironically, consumers apply sunscreen as an anti-aging treatment. And remember, doctors have been endorsing conventional sunscreen for years.

Our health is a direct reflection of the planet, and vice versa. By jumping on the "crunchy train," which means avoiding toxic chemicals and using sustainable products, the planet will benefit, too. Anything we release into the environment, be it gasoline, industrial chemicals, or heavy metals, does not evaporate or leave the planet. These pollutants enter our water. Our food. Our bodies. Our children.

In fact, 21 billion pounds of toxic chemicals are released into our environment annually.[6] In 2009, the Environmental Working Group (EWG), a leading environmental non-profit organization, tested the umbilical cords of ten babies. Researchers found in them Bisphenol A (BPA), a chemical found in plastics; tetrabromobisphenol A (TBBPA), found in computer circuit boards; synthetic fragrances from cosmetics; and perfluorobutanoic acid, a chemical related to Teflon.[7]

---

5   Group, Edward. "5 Dangerous Chemicals in Sunscreen." Global Healing Center. Global Healing Center, October 21, 2015.

6   "Toxic Chemicals Released by Industries This Year, Tons." Worldometers. Worldometers, n.d.

7   Scheer, Roddy, and Doug Moss. "Why Are Trace Chemicals Showing Up in Umbilical Cord Blood?" Scientific American. Scientific American, September 1, 2012.

However, the degree to which you are affected by these chemicals, and others, depends upon where you live, your genetics, your home, your lifestyle, your profession, and your immune system. Despite individual variations, more than half of Americans have a chronic illness, drug problem, or mental illness.[8]

By living healthier lives, we are helping our fellow earth dwellers and the planet, too. For environmentalists, focusing on lifestyle is a good initial step in helping the earth. Conversely, the health of those focused on sustainability will also benefit. Start from either side and your approach is a win-win. I dream of a world where all the products on store shelves are safe, healthy, and sustainable. I dream of a world where all restaurants serve organic food. I dream of a world where the umbilical cords of newborn babies do not contain any toxic chemicals.

THE PURPOSE BEHIND THIS BOOK

I hope that after reading this book, you will think twice about your habits and daily routines. If this book motivates you to use a sustainable glass water bottle instead of purchasing one made out of disposable plastic, I will consider this book a success. If you think twice about using carcinogenic

---

8   Mercola, Dr. "More Than Half of Americans Have Chronic Illnesses." Mercola.com. Mercola, November 30, 2016.

make-up that has had no prior safety testing, I'll be thrilled. If you wear an organic, sustainable t-shirt, I'll be delighted. I want living naturally to be a permanent trend.

I'm not insisting that you throw away (or compost) all your possessions, move to an all-organic farm, and never eat junk food again. However, if this book inspires you to do that, props to you. Live your best life. Awareness is the first step, and habit changes add up.

We live on a capitalist-driven planet. We vote with our dollars. Buying products is a direct message to companies. If natural products are flying off the shelves, then other companies will be forced to step up their game. In fact, they already have. Anna Lappe, author and sustainability activist, says it best:

"Every time you spend money, you're casting a vote for the kind of world you want."[9]

While Lappe is correct, healthier and more sustainable products are often more expensive and unattainable for some people. While this book does showcase some pricier brands I admire and hope will inspire you, I will also be highlighting cheaper alternatives whenever possible. In addition, at the

---

9   Lappé, Anna. "Anna Lappé > Quotes > Quotable Quote."
     Goodreads. Goodreads, n.d.

end of each chapter, I'll be including steps you can take to protect yourself and the planet.

Last but not least, I should conclude by thanking my mother for trying to protect me from toxins, chemicals, and unhealthy practices.

I get it now.

PART 1

OVERVIEW

CHAPTER 1

THE PRECAUTIONARY PRINCIPLE

———

The precautionary principle is the backbone of this book. The definition is: "When an activity raises threats of harm to human health or the environment, precautionary measures should be taken, even if some cause-and-effect relationships are not fully established scientifically."[10] Why not just avoid buying products and using chemicals that *might* be dangerous, just in case? Better to do that than expose ourselves voluntarily and regret it later.

Humans are an innovative bunch. We are always inventing products that make life more convenient and enjoyable. With

———
10   "Precautionary Principle." ScienceDirect. Elsevier B.V. , 2018.

the Internet and globalization, people have access to products from all over the world. However, in a free market system, capitalism and unbiased science can butt heads. Plus, science is always evolving and sometimes only later do we realize something that was meant to help our lives actually harmed our health and the environment.

Here are a few examples.

MERCURY ANTIDEPRESSANTS

The United States' sixteenth president, Abraham Lincoln, took antidepressants known as "blue mass" tablets, which doctors frequently prescribed during the nineteenth century. What are the ingredients of a "blue mass" tablet? Just 9,000 times the amount of mercury that is considered safe today.[11]

When taking the pills, Lincoln exhibited rage. However, once he stopped taking the tablets, he became visibly calmer.[12]

Now this type of treatment sounds ludicrous. However, mercury is actually *still* used—in amalgam fillings in dentistry.

---

11   Duckworth, Lorna. "Scientists Link Abraham Lincoln's Fits of Rage to Mercury Poisoning." The Independent. Independent Digital News and Media, July 18, 2001.

12   Ibid.

Each amalgam filling is made up of about 50 percent mercury.[13]

The FDA concludes that mercury in amalgams is safe for anyone over the age of six:

> FDA considers dental amalgam fillings safe for adults and children ages 6 and above. The weight of credible scientific evidence reviewed by FDA does not establish an association between dental amalgam use and adverse health effects in the general population. Clinical studies in adults and children ages 6 and above have found no link between dental amalgam fillings and health problems.[14]

However, The International Academy of Oral Medicine and Toxicology (IAOMT) counters the FDA's conclusion, asserting that mercury fillings could contribute to the development of Alzheimer's disease, Parkinson's disease, autism, anxiety, and depression.[15]

---

13 Miriam Varkey, Indu, Rajmohan Shetty, and Amitha Hegde. "Mercury Exposure Levels in Children with Dental Amalgam Fillings." International Journal of Clinical Pediatric Dentistry. Jaypee Brothers Medical Publishers, February 9, 2015.

14 "About Dental Amalgam Fillings." U.S. Food and Drug Administration. FDA, n.d.

15 "Dental Amalgam Mercury Fillings and Danger to Human Health." IAOMT. The International Academy of Oral Medicine & Toxicology, 2016.

SMOKING

In 2019, we all know that cigarettes are carcinogenic. However, while scientists began investigating the connection between ill health effects and smoking in the 1920s, they did not confirm until 1954 that smokers had higher death rates.[16]

Disturbingly, doctors *endorsed* cigarettes publicly beginning in the 1930s.[17] For example, a Camel cigarette television commercial from 1949 features an actor dressed as a doctor in a lab coat smoking as he looks over medical paperwork, and an authoritative voice-over begins:

> In a repeated national survey, doctors in all branches of medicine, doctors in all parts of the country, were asked, "What cigarette do you smoke, doctor?" Once again, the brand named most was Camel. Yes, according to this repeated nationwide survey, more doctors smoked Camels than any other cigarette.[18]

16   Mendes, Elizabeth. "The Study That Helped Spur the U.S. Stop-Smoking Movement." American Cancer Society. American Cancer Society, January 9, 2014.

17   Little, Becky. "When Cigarette Companies Used Doctors to Push Smoking." History.com. A&E Television Networks, September 13, 2018.

18   *More Doctors Smoke Camels Than Any Other Cigarette. YouTube.* YouTube, 2006.

When the truth about the dangers of smoking emerged in the 1950s, Big Tobacco had to develop a new strategy if they wanted to remain successful. They partnered with John W. Hill, the president of a public relations firm, Hill+Knowlton. Hill encouraged tobacco companies to publicly support scientific discourse.

According to Allan M. Brandt, medical historian and professor at Harvard University:

> [Hill's] strategy for ending what the tobacco CEOs called the hysteria linking smoking to cancer was to insist that there were two sides in a highly contentious scientific debate . . . . Hill would engineer controversy. This strategy— invented by Hill in the context of his work for the tobacco industry—would ultimately become the cornerstone of a large range of efforts to distort scientific process for commercial ends during the second half of the 20th century.[19]

Hill and Big Tobacco labeled as propaganda research proclaiming the dangers of smoking, and they worked on collecting public statements from scientists and doctors who

---

19   Brandt, Allan M. "Inventing Conflicts of Interest: A History of Tobacco Industry Tactics." American Journal of Public Health. PMC, January 2012.

were also skeptical. Many of these skeptics smoked cigarettes.[20] Despite protecting the reputation of smoking at the time, the tobacco industry has since lost billions of dollars from countless lawsuits against it.[21]

FLAME RETARDANTS

Beginning in the 1970s, manufacturers of furniture, electronics, toys, and even clothing treated their products with flame retardants to prevent fires. Flame retardants are chemicals now linked to cancer, infertility, and neurological issues.[22]

Flame retardants have even negatively affected firefighters. Before flame retardants entered the market, when firefighters would do their work, the burning materials were more natural and did not make people as sick. Today, because of flame retardants and other toxic chemicals, firefighters are suffering. In a video produced by the National Resources Defense Council (NDRC), an environmental non-profit advocacy group, Dr. Sarah Janssen, MD, describes the dangers of flame retardants:

---

20 Ibid.
21 Michon, Kathleen. "Tobacco Litigation: History & Recent Developments." NOLO. NOLO, 2019.
22 Callahan, Patricia, and Sam Roe. "Flame Retardants: A Dangerous Lie." The Telegraph. The Telegraph, May 26, 2012.

Flame retardants get into our bodies because they evaporate really slowly from the products that they're in, and they attach to dust particles. And we can either breathe those in, or we can ingest them by touching our hands and then touching our faces or our mouths with them.[23]

According to the International Association of Fire Fighters, 60 percent of firefighter deaths are from cancer. The scientific consensus is that these cancers are at least partially caused by the amount of chemicals to which firefighters are exposed on a daily basis.[24]

Now that research supports the notion that flame retardants are dangerous to the health of humans and wildlife, individual states have adopted flame retardant-related policies. For example, New York prohibits the sale of organohalogen flame retardants in children's toys, while Hawaii prohibits the sale of products that contain more than 0.1 percent of pentaBDE or octaBDE.[25]

---

23 "Firefighter Calls for Action on Toxic Flame Retardant Chemicals." NRDC. June 11, 2019. Accessed June 23, 2019.

24 McKay, Jim. "Firefighters Turn to Chemical Detox Saunas to Thwart the Cancer Threat." Government Technology State & Local Articles - E.Republic. April 3, 2018. Accessed June 23, 2019.

25 "Toxic Flame Retardants." Safer States. Accessed June 10, 2019.

STEPS FOR CONSCIOUS CONSUMERS

1. On the hunt for children's pajamas? Look for pajamas labeled "snug fit." This label means the pajamas were not sprayed with flame retardants. A permanent label should reveal that the item is not flame resistant, in addition to a disposable yellow hang tag restating the same information. Pajamas for infants under nine months are not sprayed with flame retardants, so a snug-fit label is not needed.[26]

2. Ashley Furniture, Crate and Barrel, IKEA, and Ethan Allen are four furniture companies that have discontinued the use of flame retardants in their furniture.[27] You can find "clean" furniture advice on the Environmental Working Group (EWG) website. Just search EWG's Healthy Living: Home Guide and click on the section regarding upholstered furniture. Also see Chapter 3 for more complete information about the EWG.

---

26 Saltzburg, Tara. "5 Things To Look For When Buying Children's Sleepwear." Westyn Baby, n.d.

27 King, Delaney. "You Can Find Hundreds (!) of Couches Without Toxic Flame Retardants." EWG. EWG, August 31, 2016.

CHAPTER 2

APPLYING THE PRECAUTIONARY PRINCIPLE TO CURRENT TECHNOLOGY

—

We are using technology at an unprecedented rate, contributing to toxicity issues worldwide by discarding broken and used technological devices and equipment into landfills. Sadly, electronics are full of heavy metals and flame retardants.[28] Our voracious appetite for technology is also creating a different form of pollution—one we can't see and many

---

28 Gross, Terry. "After Dump, What Happens To Electronic Waste?" NPR. NPR, December 21, 2010.

of us cannot feel or do not realize is affecting our overall health: electromagnetic fields, also known as EMFs.

EMFs are created by electricity in motion. Two kinds of EMFs exist—"ionizing radiation" and "non-ionizing radiation." Ionizing radiation includes obviously damaging energies, such as X-rays and gamma rays. Non-ionizing radiation, which includes radiofrequency radiation (RF), is a type of radiation emitted by cell phones, WiFi, smart meters, Bluetooth, and wireless technologies, such as the Alexa device.[29]

Because the average person uses multiple wireless devices, humans are bombarded by RF radiation daily in their homes, at work, in school, and outside. In *Generation Zapped*, a documentary about the dangers of EMFs, an associate professor of neuroscience at Karolinska Institute in Sweden, Olle Johansson, says, "If I ask you how much more such radiation does penetrate your body today compared to ten years ago, is it twice as much? Three times as much? No, it's a quintillion-times more. That's a one with eighteen zeros."[30] We are swimming in EMFs all day.

The scientific consensus is that non-ionizing radiation is not dangerous, as it cannot instantly harm us and destroy our DNA the way ionizing radiation can. However, other

---

29 Pineault, Nick. *The Non-Tinfoil Guide to EMFs: How To Fix Our Stupid Use of Technology*, N&G Média Inc., 2017, p. 5.
30 *Generation Zapped*, 2017.

scientists have found potential negative effects of wireless technologies that occur over time.[31] In fact, in 2011, the International Agency for Research on Cancer, a division of the World Health Organization (WHO), classified EMFs emitted by wireless devices as a class 2B carcinogen.[32]

Nick Pineault, digital journalist and author of *The Non-Tinfoil Guide to EMFs*, emphasizes in our interview that exposure to EMF radiation varies, depending upon location and personal habits. He says of North America, "We are [dramatically] more permissive when it comes to the amount of maximum EMF radiation that is allowed to be emitted by a cell phone tower, cell phones, or technology, compared to Italy, Belgium, China, or Russia." Why is that? President Bill Clinton signed the Telecommunications Act of 1996. This law prevented local and state governments from limiting EMF emissions as long as telecommunication companies adhered to Federal Communications Commission (FCC) regulations.[33] Those regulations generally supported technology companies and boosted the economy, but ultimately failed to protect the health of the consumer.

---

31 Pineault, Nick. *The Non-Tinfoil Guide to EMFs: How To Fix Our Stupid Use of Technology*, N&G Média Inc., 2017, p. 34.
32 Hardell, Lennart. "World Health Organization, Radiofrequency Radiation and Health - a Hard Nut to Crack (Review)." International Journal of Oncology. D.A. Spandidos, August 2017.
33 Degnan, Peter M., Scott A. McLaren, and Michael T. Tennant. " Telecommunications Act of 1996: 704 of the Act and Protections Afforded the Telecommunications Provider in the Facilities Sitting Context, The." Michigan Law. Michigan Law, 1997.

What's so wrong with RF radiation in the first place? Nick says that hundreds of studies have shown that oxidative and DNA damage occur on a cellular level when we are exposed to RF radiation.[34][35] People can get symptoms such as headaches,[36] fatigue, sleep disturbances and insomnia,[37] sperm quality and count reduction,[38] infertility,[39] depression, anxiety,[40] and more.[41]

---

34 Dasdag, Suleyman, and Mehmet Zulkuf Akdag. "The Link between Radiofrequencies Emitted from Wireless Technologies and Oxidative Stress." Journal of Chemical Neuroanatomy. ScienceDirect, September 12, 2015.

35 Lai, Henry. "Neurological Effects of Radiofrequency Electromagnetic Radiation Relating to Wireless Communication Technology." The EMR Policy Institute . University of Washington, 1997.

36 Redmayne, Mary, Euan Smith, and Michael J. Abramson. "The Relationship between Adolescents' Well-Being and Their Wireless Phone Use: a Cross-Sectional Study." Environmental Health. BioMed Central, October 22, 2013.

37 Pall, Martin L. "Microwave Frequency Electromagnetic Fields (EMFs) Produce Widespread Neuropsychiatric Effects Including Depression." Journal of Chemical Neuroanatomy. ScienceDirect, August 21, 2015.

38 Yildirim, Mehmet Erol, Mehmet Kaynar, Huseyin Badem, Mucahıt Cavis, Omer Faruk Karatas, and Ersın Cimentepe. "What Is Harmful for Male Fertility: Cell Phone or the Wireless Internet?" The Kaohsiung Journal of Medical Sciences. ScienceDirect, July 26, 2015.

39 Behari, Jitendra, and Paulraj Rajamani. "Electromagnetic Field Exposure Effects (ELF-EMF and RFR) on Fertility and Reproduction." Semantic Scholar. Semantic Scholar, November 2012.

40 Keller-Byrne, Jane E., and Farhang Farhang Akbar-Khanzadeh. "Potential Emotional and Cognitive Disorders Associated with Exposure to EMFs." Sage Journals. Sage Journals, February 1997.

41 Havas, Magda. "Radiation from Wireless Technology Affects the Blood, the Heart, and the Autonomic Nervous system1)." Reviews on Environmental Health. De Gruyter, November 5, 2013.

RF radiation has even been shown to cause cancer in both rats and humans. In 2016, the US National Toxicology Program (NTP) applied RF radiation to rats and mice equivalent to the amount of radiation a human would experience if he or she talked for thirty minutes a day on a cell phone for thirty-six years. The results? The RF-exposed rats had higher rates of two kinds of cancer, glioma and malignant schwannoma of the heart, than the control group.[42] In 2013, a case study was published on PubMed, in which four women were documented as storing their phones for hours a day in their bras for multiple years before being diagnosed with breast cancer.[43]

A percentage of the population has EHS, which stands for electrohypersensitivity. These people feel acute physical and mental symptoms when exposed to RF radiation, and sometimes they can be completely disabled.[44] Dr. William Rea, MD, in Dallas, Texas, conducted a study of EHS sufferers in 1991. The majority of the patients were able to feel when RF

---

42 Pineault, Nick. *The Non-Tinfoil Guide to EMFs: How To Fix Our Stupid Use of Technology*, N&G Média Inc., 2017, p. 35.

43 West, John G, Nimmi S Kapoor, Shu-Yuan Liao, June W Chen, Lisa Bailey, and Robert A Nagourney. "Multifocal Breast Cancer in Young Women with Prolonged Contact Between Their Breasts and Their Cellular Phones." Case Reports in Medicine. Hindawi Publishing Corporation, 2013.

44 "Electrohypersensitivity Overview." Physicians for Safe Technology. Physicians for Safe Technology, September 26, 2017.

radiation was being directed at them and when it was not.[45] Just as some people are more noticeably affected by bad air quality or mold, certain people will notice the immediate effects of EMFs while others might not be aware.

If RF radiation is so bad, why are more people not suffering? While many people are ill and EMFs could play a role in their health pictures, the effects of RF exposure are *cumulative*, meaning that we cannot easily pinpoint EMFs as being the sole cause of health issues. After all, the average person encounters a lot more pollution than just EMFs, and long-term studies on the effects of manmade EMFs have not been conducted.

WHAT IS 5G?

You may have some awareness that 5G is coming, which is the fifth generation of wireless technologies. Ignoring the dangers, 5G actually sounds amazing—faster data transmission speeds than ever before and the ability for autonomous cars to drive themselves without human supervision. Of 5G, Nick Pineault says: "It is being installed everywhere. Basically, there are a few ways it changes technology in our

---

45   Rea, William J, Yaqin Pan, Ervin J. Yenyves, Iehiko Sujisawa, Hideo Suyama, Nasrola Samadi, and Gerald H. Ross. "Electromagnetic Field Sensitivity Case Study Evaluation." Journal of Bioelectricity, 1991.

EMF-polluted cities. The number one way is antenna inten-sification, so we are going to multiply the number of anten-nae required by multiple folds." Nick says that in a city like Montreal, Canada, where he lives, there are currently around 1,200 antennae for 4G, the current generation of wireless access. In a 5G Montreal, at least 60,000 antennae will be installed. He continues:

> It's a high-power but short-distance technology, so instead of having one or multiple antennae every square mile, you have hundreds of them. You have one every three to twelve homes in res-idential areas, and you have one or multiple new antennae every block in downtown areas. It's a highly directional exposure, so the cell tower is literally pointing toward your phone and emit-ting the radiation toward your phone.

Nick shares with me that in 2019 in Turin, Italy, the local government has indefinitely shut down an area of a park because the EMF levels emitted from cell phone towers were too high.[46] He says of the levels found in the park, "It's equiv-alent to what people in North America will be exposed to 24/7 if you live near a 5G antenna, and it's already being installed in thirty-plus US cities." Nick adds that the Minister of

46 Rambaldi, Massimiliano. "L'elettrosmog Fa Chiudere Il Vecchio Parco Giochi - La Stampa." La Stampa. La Stampa, April 26, 2019.

Environment in Brussels, Belgium, Céline Fremault, halted 5G for now, saying that she would not allow her citizens to be guinea pigs and risk their health.

Humans are not the only ones being affected by EMFs. Nick says, "The studies on nature are even stronger than in humans. You don't have a placebo effect. You just have insects or birds losing their sense of where is north and where they should migrate. You see a lot of confusion in birds, in bees, because they lose their innate sense of magnetism." Nick is right. In 2018, Eklipse, a European Union-funded, environmentally-focused organization, reviewed over ninety-seven EMF-related studies and concluded that RF radiation could negatively affect bird and insect populations and interfere with plant development.[47]

When it comes to wireless technology, the precautionary principle needs to be applied. Only time will tell the true effects of wireless technology, but enough conflicting evidence exists to merit preventive measures. Just as it took time for doctors to question mercury antidepressants, cigarettes, and flame retardants, learning the truth about EMFs will not happen overnight. In the interim, despite the extraordinary opportunities and economic benefits that come from

---

47   Dovey, Dana. "A Switch to 5G May Be Bad for the Environment." Newsweek. Newsweek, August 27, 2018.

fast technology, we need to pay attention to the health and environmental effects of EMFs.

STEPS FOR CONSCIOUS CONSUMERS

1. Distance is key when it comes to using technology intelligently. Never put a cell phone against your head—always use it on speaker mode or with low-EMF headphones, such as the ones that Defender Shield makes called EMF Radiation-Free Air Tube Stereo Headphones. If using your phone on speaker, one to two feet of distance from your body is ideal. The farther away you put a device, the less you are exposed to EMFs.

2. Try to avoid using your cell phone in places with poorer reception, because your phone has to send stronger signals to connect to a cell phone tower. Therefore, you'll be exposed to higher levels of RF radiation. Avoid using your phone in parking garages, cars, elevators, and planes (unless you're in airplane mode, of course).

3. Put your phone in airplane mode as much as possible, especially when sleeping. You can still use many of the apps while in airplane mode. Download apps such as Google Maps or Here We go for navigation. These apps let you navigate offline, meaning you don't need a signal to use the app. You can download podcast episodes

and Spotify playlists ahead of time and enjoy them in airplane mode. Check out my YouTube channel for a tutorial about how to hardwire your phone, so you can still use all the apps while minimizing your exposure to EMFs. All you need is a few gadgets, and this setup allows you to still make WhatsApp and Skype calls and texts.

4. Turn your WiFi router off at night. Many people report better sleep, less insomnia and anxiety, and relief from additional symptoms. Because you don't surf the Internet in your sleep, having it at night is unnecessary. Give it a try and see how you feel. If you're nervous about having no cell phone access at night, consider reinstalling a landline. (Remember those?)

5. If you want to completely eliminate your EMF home exposure, which is ideal, disable your WiFi completely and use a hardwired connection instead, using ethernet cables. Doing so is inexpensive and easy. Check out my YouTube channel for a tutorial.

6. If you would like to make sure your home is as low EMF as possible, consider investing in your own EMF meter, which evaluates the radiation in your environment, or hire a building biologist or Geovital consultant in your area who can assess your space for you.

7. Nick does not recommend using wireless anything—this includes earbuds, Bluetooth devices, "smart" devices, such as TVs (good news—you can probably hardwire yours like I do), cordless phones, and smart meters (in many places, you can opt out).

8. While we should all mitigate our EMF exposure, children should be protected at all costs. Children are much more susceptible to the negative effects of EMFs, because children's bodies contain more water than those of adults. If a mother and her child both talk on their cell phones for the same amount of time and at the same distance, the child's head will absorb double the amount of radiation and the child's bone marrow will absorb ten times as much.[48] Other countries are beginning to respond to the danger that EMFs present to children. For example, since 2015, France has banned WiFi in French nursery schools and only permits elementary schools to turn on WiFi when needed.[49]

---

48 Pineault, Nick. *The Non-Tinfoil Guide to EMFs: How To Fix Our Stupid Use of Technology*, N&G Média Inc., 2017, p. 30.

49 "France: New National Law Bans WIFI in Nursery School!" Environmental Health Trust. Environmental Health Trust, October 28, 2015.

CHAPTER 3

REGULATORY AND ADVOCACY ORGANIZATIONS

As conscious consumers, can we rely on the government or private organizations to protect us and the planet? The answer is complex.

GOVERNMENT ORGANIZATIONS

THE FOOD AND DRUG ADMINISTRATION (FDA)

Governmental organizations *are* in place to protect Americans—the most obvious of these being the Food and Drug

Administration (FDA), an agency of the federal government. The FDA states on its website:

> The Food and Drug Administration is responsible for protecting the public health by ensuring the safety, efficacy, and security of human and veterinary drugs, biological products, and medical devices; and by ensuring the safety of our nation's food supply, cosmetics, and products that emit radiation.[50]

The FDA has been in existence since 1906,[51] and as of 2018, has over 17,000 employees. While the FDA is mostly comprised of civilian employees, a small percentage is from the military.[52]

Many people have questioned the FDA's trustworthiness. Pharmaceutical companies give hundreds of millions of dollars to the FDA each year. This money goes toward speeding up pharmaceutical approvals—in fact, the FDA has the fastest review period of any regulatory agency in the world.[53] While drugs can transform and save lives, can a federal

---

50 "What We Do." U.S. Food and Drug Administration. FDA, n.d.
51 "When and Why Was FDA Formed?" U.S. Food and Drug Administration. FDA, n.d.
52 "DETAIL OF FULL-TIME EQUIVALENT EMPLOYMENT (FTE)." FDA. FDA, 2018.
53 Chen, Caroline. "FDA Repays Industry by Rushing Risky Drugs to Market." ProPublica, March 9, 2019.

agency remain neutral with financial backing from pharmaceutical companies that want their drugs to be approved quickly? The expedited process could have dangerous repercussions—thousands of deaths occur each year from adverse reactions to medications.[54] Perhaps if the review process were conducted without financial backing and at a slower rate, some of these deaths could be prevented.

Also, the FDA does not screen the ingredients in cosmetics unless they contain previously banned chemicals.[55] Cosmetics include make-up, moisturizers, hair products, perfume, and nail products.[56] While the FDA has only banned eleven chemicals in cosmetics, Europe has banned over 1,300.[57] The FDA requires cosmetic companies to list ingredients, but allows brands to use the vague identifier "fragrance," rather than disclose individual chemicals.[58] Why? In 1966, the government enacted a law called the Fair Packaging and Labeling Act. The Act states that companies do not have to disclose

---

54 Light, Donald W. "Risky Drugs: Why The FDA Cannot Be Trusted." Edmond J. Safra Center for Ethics. Harvard University, July 17, 2013.

55 "Cosmetics Safety Q&A: Prohibited Ingredients." U.S. Food and Drug Administration. FDA, n.d.

56 "Small Businesses & Homemade Cosmetics: Fact Sheet." U.S. Food and Drug Administration. FDA, n.d.

57 Milman, Oliver. "US Cosmetics Are Full of Chemicals Banned by Europe – Why?" The Guardian. Guardian News and Media, May 22, 2019.

58 "Fragrances in Cosmetics." U.S. Food and Drug Administration. FDA, n.d.

the ingredients in their signature fragrances because they are "trade secrets."

According to the Environmental Working Group (EWG), on average, a fragrance used in a perfume contains fourteen different chemicals. These chemicals can include endocrine-disrupting phthalates, potential carcinogens, and allergens. The majority of these chemicals have not been tested thoroughly for human safety.[59] Undisclosed fragrances can also be found in cleaning products, candles,[60] and diapers.[61]

In our interview, Jeffrey Hollender, co-founder of Seventh Generation, Inc. and current CEO and founder of The American Sustainability Business Council (ASBC), expresses concern about the FDA and disclosure of fragrance ingredients:

> The smells of certain cleaning products that we experience as children, we often begin to identify with what it means to be clean, and unfortunately, many of those scents may mean clean but should also mean dangerous and highly toxic.

---

59 Schrock, Monica. "WTF Is In Fragrance and Is It Harmful!?" Non Toxic Revolution. Non Toxic Revolution, May 30, 2017.

60 Frack, Lisa, and Becky Sutton. "3,163 Ingredients Hide Behind the Word 'Fragrance.'" EWG. EWG, February 2, 2010.

61 "Non-Toxic Diapers: Safer Disposable Diapers for Babies." The Gentle Nursery . The Gentle Nursery , n.d.

He continues that these scents often contain volatile organic compounds (VOCs), which trigger asthma and allergies. He adds, "The answer from a consumer's perspective is, if they won't tell you what's in the fragrance, then don't buy it, because you're putting your health at risk." On a hopeful note, Jeffrey believes that fragrance transparency has reached a tipping point. SC Johnson and Unilever, two personal care product corporations, have recently promised to provide better fragrance disclosure.

The FDA also is not concerned about glyphosate, a well-known and potentially carcinogenic herbicide. The agency states on its website:

> The EPA [Environmental Protection Agency] evaluates the safety of pesticides such as glyphosate. According to [the] EPA, glyphosate has a low toxicity for people. Pets may be at risk of digestive or intestinal problems if they touch or eat plants that have just been sprayed.[62]

The Monsanto Company, which is now owned by the German company, Bayer, sells a weed-killing product called Roundup. Roundup contains glyphosate. Roundup is used by both individual consumers for home gardens, as well as by farmers, who spray the product on certain crops to eradicate difficult

---

62 "Questions and Answers on Glyphosate." U.S. Food and Drug Administration. FDA, n.d.

weeds or to increase harvest.[63] As of 2019, over 13,000 people have filed lawsuits against Bayer, claiming that Roundup caused them to develop non-Hodgkin's lymphoma.[64]

THE CONSUMER PRODUCT SAFETY COMMISSION (CPSC)

Established in 1972, CPSC's mission is to "protect the public against unreasonable risks of injury or death from consumer products through education, safety standards activities, regulation, and enforcement."[65] The Commission is considerably smaller than the FDA, yet had over 500 employees as of 2012. The CPSC regulates categories such as clothing, toys, electronics, furniture, and lighting. The CPSC website lists recalled items. For example, a trampoline that collapsed too easily was recently recalled,[66] in addition to children's pajamas that did not meet the flammability standards for children's clothing,[67] and an essential oil that lacked proper child safety packaging.[68]

63 "Glyphosate." Wikipedia. June 07, 2019. Accessed June 10, 2019.
64 "Monsanto Papers." U.S. Right to Know. USRTK, n.d.
65 "Frequently Asked Questions and Answers (FAQs)." CPSC.gov. CPSC, n.d.
66 "Super Jumper Recalls Trampolines Due to Fall and Injury Hazards." U.S. Consumer Product Safety Commission, August 1, 2019.
67 "H&M Recalls Children's Pajamas Due to Violation of Federal Flammability Standard." U.S. Consumer Product Safety Commission, July 25, 2019.
68 "Wintergreen Essential Oil Recalled by Epic Business Services Due to Failure to Meet Child Resistant Closure Requirement; Risk of Poisoning (Recall Alert)." U.S. Consumer Product Safety Commission, July 16, 2019.

The CPSC focuses on obvious mishaps that could result from using a product. For example, children's pajamas sized nine months to fourteen years must either be snug fitting or pass flammability tests.[69] To pass flammability tests, the manufacturer must use flame retardants.[70] As discussed previously, flame retardants have health risks.

THE FEDERAL TRADE COMMISSION (FTC)

Established in 1914, this agency with more than 1,000 employees[71] is more economy-focused. The website states the organization's mission: "Protecting consumers and competition by preventing anticompetitive, deceptive, and unfair business practices through law enforcement, advocacy, and education without unduly burdening legitimate business activity."[72] The agency prohibits anticompetitive business behavior.[73] The agency also protects the consumer by establishing the Do Not Call Registry,[74] preventing identity theft,[75] promoting honesty in marketing,[76]

---

69 "Children's Sleepwear Regulations." CPSC.gov. CPSC, April 10, 2019.
70 Saltzburg, Tara. "Snug Fit Pajamas: A Guide to Flame Retardants and the Children's Sleepwear Regulations." Westyn Baby, February 14, 2019.
71 "Careers at the FTC." Federal Trade Commission, n.d.
72 "About the FTC." Federal Trade Commission. FTC, n.d.
73 "Mergers and Competition." Federal Trade Commission. FTC, n.d.
74 "The Do Not Call Registry." Federal Trade Commission. FTC, n.d.
75 "Identity Theft." Federal Trade Commission. FTC, n.d.
76 "Truth In Advertising." Federal Trade Commission. FTC, n.d.

and keeping technology companies accountable and transparent with their customers.[77]

NOTABLE NON-GOVERNMENT ORGANIZATIONS

ENVIRONMENTAL WORKING GROUP (EWG)

In 1993, concerned environmental activists Ken Cook and Richard Wiles established the Environmental Working Group (EWG), a non-profit, non-partisan organization. The group is comprised of scientists, policy experts, lawyers, and computer programmers who focus on protecting consumers by educating the public about toxic chemicals in everyday products. EWG workers are the superheroes of conscious consumerism—revealing the truth even when it interferes with sales. The group regularly releases data on such topics as glyphosate levels in food products sold in grocery stores, pollutants found in the water supply, and chemicals in mainstream sunscreens.

Ken Cook, the EWG's co-founder and current president, believes that the EWG is essential in protecting the American consumer. In a video commemorating the twentieth anniversary of the organization, Ken states that EWG's mission is to turn smart consumers into advocates.[78]

---

77 "Mobile Technology Issues." Federal Trade Commission, n.d.
78 Cook, Ken. "Ken Cook on EWG's 20th Anniversary." YouTube. YouTube, October 28, 2013.

Katherine Baird, the Special Projects Manager of the EWG, agrees that consumer protection is the group's central motivation, telling me in an interview:

> Our work falls into several categories. Our science teams perform in-depth investigations to identify contaminants and exposure routes, to better understand our "exposome," the range of exposures one encounters in a day.

This team also runs the various EWG databases, such as Skin Deep, Food Scores, and a Guide to Healthy Cleaning, which are meant to help the average consumer make smart, healthy, and sustainable decisions. She continues:

> Our agriculture team works to map and identify environmental threats like regions with PFAS-contaminated water, or areas of the country where toxic algae blooms are popping up. Our government affairs team works tirelessly on the Hill here in D.C., as well as in California, to develop new laws and regulations that aim to protect consumers. Lastly, our digital team works to spread the word and make our guides and tip sheets available and accessible to everyone.

The EWG staff includes a team of scientists who specialize in chemistry and toxicology, as well as professionals with Masters degrees in population health science. The EWG also utilizes third-party labs to do testing for certain reports, as was the case when they recently tested oat-based cereals to see how much glyphosate, a likely carcinogenic pesticide, was present. She says, "We develop the methodology for the study, collect the samples, and then send them to the lab (usually blind), and then analyze the results in-house."

She shares with me that the average American consumer believes that if a product is on a store shelf, it is automatically safe, which sadly isn't the case. She says:

> Right now our federal regulations aren't built in a way that genuinely protect consumers. Our cosmetics law hasn't been updated since 1938, and when the law was originally written, it built a self-policing system for the cosmetics industry. Just recently, tests found asbestos [a known carcinogen] in children's cosmetics sold at Claire's stores. The FDA had no power to compel Claire's to recall the products or halt their sales, and the only immediate action the FDA could take was to write a letter [to the company].

The EWG *is* making a difference. New companies and retailers are setting higher, safer standards for products. Each year, the EWG website gets over 25 million visitors and the Skin Deep database, developed by the EWG, gets 8.8 million searches. Consumers can use the Skin Deep database to search for products and individual ingredients to determine what the EWG recommends.

Because of higher demand for transparency, consumers can now easily find information about companies and products—both good and bad. Katherine elaborates:

> Major brands, some with less than sterling histories, are losing their market share to smaller brands that are doing it right, and they're responding. There is major work to be done. Our hope is that we will have regulations in place that enforce the kind of product safety that we deserve, so that truly, every product that makes it to a store shelf or an online marketplace will be safer to use—so until we get there, there is a lot of work to do.

She adds that the Internet has helped people inform themselves about smart consumer practices, which has encouraged new companies to form and people who are passionate about consumer protection to step forward and help change policies.

I ask her what the EWG would like to see from the FDA. She says:

> The FDA (and EPA and USDA) should be setting stricter standards for chemicals in our environment. We are desperately far behind in our regulations. A major piece of environmental legislation hasn't been passed in almost thirty years. We haven't set a new drinking water standard in almost twenty years. We're working our hardest to change things on Capitol Hill, but we can't wait for them.

She continues that the EWG has noticed significant shifts in the marketplace. For example, the demand for organic food is higher than ever in America and is so much more accessible than it was in the 1990s. She adds:

> Same with personal care and cleaning products—the breadth of healthier, safer products that are available now is stunning. You are seeing major retailers like Walmart and CVS making commitments to reduce the chemicals of concern used in their products. This is all because consumers demanded it. We are working on demonstrating the value in getting ahead of the government and not waiting for their regulation because it's what consumers want.

THE AMERICAN SUSTAINABILITY
BUSINESS COUNCIL (ASBC)

ASBC is a network of over 250,000 businesses that support pass-
ing responsible legislation that regulates and encourages businesses to create positive social change. Some of these businesses
include the ice cream maker, Ben and Jerry's; the clothing chain,
Eileen Fisher; the cosmetic and skincare brand, Beautycounter;
and the e-commerce site, Etsy. ASBC advocates for improving
ingredient transparency, raising the minimum wage, establishing better employee benefits, reducing the amount of packaging
in the marketplace, and protecting our water supply.

Jeffrey Hollender, co-founder of Seventh Generation and
CEO of ASBC, started the organization in 2012 because he
believes businesses can be doing better to create a sustainable
and healthy world. He thinks that while consumers vote with
their dollars, the responsibility of creating a greener marketplace should not be placed solely upon the consumer. He says:

> Right now, we are practicing an extreme version
> of capitalism, where we are allowing a very small
> group of people to accumulate ungodly amounts
> of money and power, and for that wealth and
> power to become incredibly concentrated in a way
> that is dangerous and disadvantageous to the poor
> and middle class. We need to have much greater
> regulation that protects the public and the planet

from the negative impacts of these companies. By externalizing those impacts, they take them away from their expenses and their responsibility.

The ASBC has an annual summit, where environmentally-focused business founders and advocates meet to discuss how capitalism can fix environmental and social issues. The ASBC's website is also full of compelling case reports, highlighting how businesses switching to more sustainable practices, such as eliminating toxic chemicals, can benefit everyone.

STEPS FOR CONSCIOUS CONSUMERS

1. Download the free EWG Healthy Living app, where you can search for personal care, cleaning, and food products by name, or you can even scan the barcode on the product itself.

2. Download the free Think Dirty app, which Lily Tse, a clean-beauty advocate, created in 2013.[79] This app is similar to the Healthy Living app but is not associated with the EWG.

3. For more information about the EWG, check out its site at www.ewg.org.

---

79 Tse, Lily. "A MESSAGE FROM THE FOUNDER." Think Dirty. Think Dirty, n.d.

CHAPTER 4

GREENWASHING

Companies are starting to realize that many consumers want to lead more sustainable and healthier lives. As much as I'd like to throw (ethically-sourced, compostable) confetti into the air as I report this information to you, some companies are taking unfair advantage of the trend toward eco-friendly and healthy products. Some businesses are manipulating consumers. They try to convince shoppers that their products are sustainable, ethical, and healthy via a marketing technique called "greenwashing." Greenwashing occurs when products are marketed with buzzwords, such as "green," "healthy," "non-toxic," "natural," and "earth-friendly," when actually they might contain carcinogens and other toxic materials.

Laura Ehlers, social media influencer and green beauty expert, tells me in an interview that we must read every ingredient label thoroughly before purchasing a product. She says of greenwashing, "I like to define it as marketing or packaging that makes a product seem natural when it isn't, or it might have a couple of natural ingredients sprinkled in there for marketing purposes, but still contains harmful chemicals."

She says even organic labeling can be deceiving. In America, if a food product's label says "organic," that under United States Department of Agriculture (USDA) guidelines, 95 percent of the ingredients must be organic. Look for the 100 percent organic label if you want to make sure a food item is exclusively organic. However, if a food product says it is "made with organic ingredients," then only 70 percent of the ingredients have to be organic.[80]

Greenwashing is a modern epidemic, and I, too, have been fooled by it. Just recently I realized that the "natural" detergent I had been using for years contained potentially problematic ingredients, including methylisothiazolinone. Researchers found in 2002 that this chemical severely harmed rats.[81] I hadn't questioned the ingredients of the detergent because I had been using it for years and assumed the brand would protect me. I ultimately switched to a more

---

80 "Organic Labeling Standards." Organic Labeling Standards | Agricultural Marketing Service. USDA, n.d.

81 "Methylisothiazolinone and Methylchloroisothiazolinone." Campaign for Safe Cosmetics, n.d.

inexpensive and local detergent brand that has all non-toxic and safe ingredients. I would have continued to be fooled by the brand's "green" marketing if I hadn't delved deeper on my own. Doing your own research (DYOR) is essential.

DYOR is not BYOB (bring your own beer), but why not crack open an organic, glyphosate-free beer as you read this book (if you're over twenty-one, of course). I encourage you not to be spoon-fed ideas by organizations, professionals, or the person in front of you in line at Starbucks. Instead, question what you have been told. Do not trust a product just because it made its way onto a shelf in a store, even a health food store. Research the ingredients in the products you buy. Investigate how these products are made. How are the workers at the factories treated? Decide which products you feel comfortable using and supporting.

STEPS FOR CONSCIOUS CONSUMERS

1. Go through your personal care and cleaning products. If you find "fragrance" on an ingredient list, do not purchase this product again. Remember, fragrance can mean anything. Laura Ehlers recommends taking a two- to four-week break from synthetic fragrance. Often times, people are surprised to learn that these products were making them sick.

2. Laura says that switching to clean beauty and non-toxic products is going to be a marathon, not a sprint. Don't

throw away all you own and try to replace everything immediately—that will be too overwhelming, expensive, and stressful. Go slowly, and replace as you go. Next time you need something, make a healthier choice.

3. Want to avoid ingesting glyphosate? While corn and soy crops are sprayed with glyphosate, surprisingly, foods that are contaminated the most with glyphosate are oat and wheat-based products. The EWG tested conventional and organic oat and wheat-based products and found that Cheerios, Quaker Oats, Kellogg's, and Nature Valley products scored high. Avoid these brands, and purchase organic instead.[82]

4. Practice identifying greenwashing at the store. Just because a product is labeled natural or healthy does not mean anything.

82 Temkin, Alexis. "Breakfast With a Dose of Roundup?" EWG. EWG, August 15, 2018.

PART 2

THE ECOPRENEURS

CHAPTER 5

CAPITALISM AND HEALTH

Capitalism and health have a weird relationship in America. The United States and New Zealand are the only two countries in the world where advertising for pharmaceuticals is allowed on television.[83] Should patients suggest drugs to their doctors simply because a commercial playing during *The Bachelorette* shows a medicated actress dancing in the sunshine?

In America, there are over 67,000 pharmacies, the majority of which are retail chains—places like Walgreens, CVS, and

---

83   Marshall, John. "Why You See Such Weird Drug Commercials on TV All the Time." Thrillist. Thrillist, March 23, 2016.

Rite Aid.[84] A study conducted by *Consumer Reports* surveyed nearly 2,000 adults and found that 55 percent frequently take a prescribed medication. These individuals take an average of four medications, in addition to over-the-counter medications and supplements.[85]

Because many patients see multiple doctors for prescriptions, dangerous adverse drug reactions commonly occur. Over a third of the people surveyed who take prescription drugs and see multiple doctors reported that their medical team had not determined whether all their drugs were medically necessary. Nevertheless, pharmaceutical consumption is on the rise.[86] In fact, America has the largest pharmaceutical industry in the world, taking in over $460 billion in revenue in 2018.[87]

THE AVERAGE AMERICAN DRUGSTORE

Imagine walking into the average pharmacy to fill your prescription. The fluorescent lighting makes you feel like you're a specimen in a lab. You walk by the cosmetics aisle and pass by

---

84 Qato, Dima Mazen, Shannon Zenk, Jocelyn Wilder, Rachel Harrington, Darrell Gaskin, and G. Caleb Alexander. "The Availability of Pharmacies in the United States: 2007–2015." PLOS ONE. Public Library of Science, August 16, 2017.

85 Preidt, Robert. "Americans Taking More Prescription Drugs Than Ever: Survey." Consumer HealthDay. HealthDay, August 3, 2017.

86 Ibid.

87 "Worldwide Pharmaceutical Sales by Region 2016-2018 | Statistic." Statista. Statista, n.d.

carcinogenic and endocrine-disrupting lip products, blushes, eyeshadows, and nail polishes. A pungent odor wafts from the cleaning aisle—a medley of toxic artificial florals, nauseating citruses, and sickeningly-sweet vanillas. You already feel ill, and then you see the line at the pharmacy counter. Your heart sinks.

Maybe you can get a snack while you wait. You make a beeline for the snack aisle, and you scan the options. Should you eat the chips that contain genetically-modified corn, Yellow 6 food coloring that potentially causes adrenal and testicular tumors,[88] and MSG? Or should you try the cookies with the high fructose corn syrup and artificial flavors? You're suddenly not hungry anymore. You quickly fill your prescription and leave, wondering why pharmacies offer these products in the first place.

PHARMACA

But wait. Someone did flip the conventional pharmacy model on its head—an alternative exists. Meet Barry Perzow, founder of Pharmaca Integrative Pharmacy, which he started nearly twenty years ago. Pharmaca, with thirty-four stores across America, is dramatically different than the average pharmacy. Yes, customers can fill their prescriptions there, but that is where the similarities end between Pharmaca and

---

88 "Summary of Studies on Food Dyes." Center for Science in the Public Interest. Center for Science in the Public Interest, n.d.

conventional drugstores. Pharmaca is staffed by healthcare professionals, such as estheticians, nutritionists, herbalists, and naturopaths, so customers have help making healthy and educated decisions.

All the products at Pharmaca are selected carefully with the health of the customer and the planet in mind. Shop at a Pharmaca, and you'll come across organic and vegan snacks, cruelty-free and chemical-free cosmetics, vitamins, and natural cleaning products. Pharmaca also offers supportive therapies. For example, if a customer needs to be on a certain pharmaceutical that might deplete a vitamin in the body, a Pharmaca representative can recommend something to offset an adverse reaction.

Barry was living an organic lifestyle before it was cool and mainstream. He started Pharmaca after spending years in the natural food industry. From 1993 to 1997, he was the president of Capers Natural Community Markets, which became the largest natural grocery chain in Canada. In 1995, Capers merged with Colorado-based food company, Alfalfa's, and the following year, merged with another food company, Wild Oats. By then, Capers Alfalfa's was 124-stores strong. The company was then sold to Whole Foods.

Barry's natural food stores had broken the conventional grocery-store model. He then decided to do the same thing for pharmacies. He saw that conventional pharmacies had room for improvement, telling me in an interview that visiting them "felt like shopping in a microwave oven. It just didn't feel good."

Barry and his wife traveled through Europe for six months, looking for a new pharmacy model. He tells me:

> I wanted to be influenced more by homeopathy and naturopathic products and less by medicated prescriptions. The average American is a quick-fix consumer who has grown into a position of saying, "If I have something wrong with me, give me the fastest thing, I don't care what it does to my body or what it does to my system, just make it go away."

He and his wife visited between 400 and 500 European pharmacies, met with the owners, and fell in love with the European pharmacy model. He admired that pharmacists in Europe preferred natural remedies, like homeopathy, to drugs. He says of the European model, "In very dire circumstances, under a doctor's advice, they will dispense a drug, but that's the last resort. In our country, that's the first resort."

He also paid particular attention to how European pharmacies were staffed with medical professionals who were actual experts in their fields, unlike American conventional pharmacies, where pharmacists were the only experts on the premises. He says:

> I started hiring only homeopathic doctors, naturopathic doctors, clinical nutritionists, wellness consultants, and estheticians. I was more interested in having an esthetician who was also a nutritionist— which we wound up finding—who would say to a customer, instead of pushing some cosmetic onto them, "Before I recommend this particular skin product, I recommend taking fish oil every day for six months, and let's get some moisture into your skin, and then I'll recommend a product."

Barry wanted the customers to feel like the staff member cared about their needs, rather than just pushing sales.

When I ask Barry about how he staffed the first store in Boulder, Colorado, he tells me that he put a whole page advertisement in the local newspaper, *The Boulder Camera*, stating that he was holding a job fair for homeopaths, naturopaths, estheticians, and those trained in Ayurvedic medicine. He did not know if he would find the people he needed. He says:

> I pull into the parking lot in my car, and I see almost fifteen people waiting in line in front of

the tent, and I say to myself, "Wow, I didn't think this was going to happen!" I start interviewing people, and I hired the first ten. And those ten people are still with us today. They were fantastic.

Barry knew he wanted to create an integrative pharmacy, which was not an established idea at the time. The unfamiliarity of an integrative pharmacy reminded him of his time working in the organic food business. He says, "When we were opening up our new stores, we were promoting ten reasons to buy organic, and people would come in and say, 'We love the concept of what you're doing, but what's organic?' People didn't even know what organic meant." Barry and his team worked on educating their customers, printing posters and bags that explained why organic products are important for health and sustainability.

Barry remembers when organic became trendy—he was watching a late-night television show, and actress Meryl Streep had an apple in her hand, explaining that she was against Alar, one of the pesticides sprayed on the majority of apples. He says:

The host asks, "What's the problem?" And she said, "Before I feed this to my child, I want to brush off, scrape off, all the herbicides and pesticides that are very dangerous." And she made a whole spiel about that. Within a week, we were having trouble keeping our rotation of fresh product in our organic produce section.

People became obsessed with organic produce, and the demand has not slowed down since.

Just as customers of his grocery stores were unfamiliar with organic food, Pharmaca's customers were initially reluctant to embrace his new integrative pharmacy model. However, after explaining what *integrative* meant, customers loved the concept and kept returning. He says:

> Integrative is where we're integrating our various types of interventions. It could be homeopathy, it could be chiropractic methods, it could be any holistic cure, like acupuncture. If a customer has a problem, it could be a combination of massage, yoga. A whole number of different interventions that come together to help keep you healthy. Therefore, integrative. Integrating all of them.

Barry shares that it took a year and a half of persistent story-telling and marketing for customers to understand that just because they are filling their prescription at an integrative pharmacy did not mean that they would be getting a different product. Pharmaca dispenses the same pharmaceuticals, but customers have additional opportunities, such as the ability to buy alternative and healthy products, and the option of chatting with medically-educated staff members.

When I ask him whether he knew he wanted to stock his store with natural and organic products all along, he says absolutely. Before opening his first store in Boulder, he spent six months designing every aspect of the store, from the shelving to the signage to the categories to the product selection:

> I created all these different zones. Naturopathic zone, homeopathic zone, a sexual intimacy product zone, way out there. People walked in and said, "We have never seen that kind of category in any kind of store before." They were pleasantly shocked, and they enjoyed the shopping experience. I actually picked every single item that went into the product mix, one by one, placed them on the shelf and figured out how I wanted to lay it out.

Barry wanted customers to trust that if a product made its way to a Pharmaca shelf, it was automatically safe.

When I ask Barry if he knew Pharmaca would be a larger chain of thirty-four stores, he tells me no, not at first. After two successful stores in Boulder, he had a meeting with his bank manager. He says, "I'm a very competitive guy. My bank manager says, 'Perzow, you know what you did is great, and I know you're having fun, and it's working, but that's such a Boulder crunchy concept. This would never work outside of Boulder.'" Barry told his bank manager that Pharmaca could be successful in many markets. His bank manager challenged him, saying that he would give him a loan and interest if Barry's pharmacies succeeded outside of Boulder. Barry says, "So that was a good challenge. I love challenges."

Barry soon opened a store in Portland, Oregon, which was a huge success almost overnight. Then he expanded to Washington State, California, and New Mexico. At first, Barry was converting already-established pharmacies, often employing the pharmacists already there. But after a while, he started opening up stores from scratch, called "greenfield stores."

When I ask him what creating Pharmaca taught him, he says:

> Maybe I make it sound easy, but believe me, any model I ever created, especially Pharmaca, was a big departure from anything that was out there at the time. When I was in the natural food business, there were many other natural food stores that had been around for a while in different parts of the country. I wasn't recreating the wheel, right? Pharmaca, I was recreating the wheel. I

was working eighteen to twenty hours a day. It takes a tremendous commitment of working hard to make sure you get it right.

He tells me that being a nonconformist, whether that involved promoting organic food when everyone eats conventional, or creating an integrative pharmacy when nothing else existed like it at the time, is difficult:

> I was an outcast when I was in my twenties and thirties—people thought I was a nutcase. The early pioneers in alternative lifestyles and alternative retail strategies, we had challenges. We stuck with our beliefs, and for many years, we starved, we couldn't make money, we lost money, we borrowed money, we raised money, but we believed in what we were doing. Over time, as you can tell today, it has paid off, because we stuck to our guns, in great distress, back in the '60s.

Barry believed that the sustainable and healthy lifestyle that he was promoting via Pharmaca was the way of the future. He is truly a health pioneer. Throughout his life, he has broken models that do not serve the planet or human health. He has challenged the status quo and shown customers across the country a different kind of health paradigm, one in which we can make educated decisions rather than blindly trusting medical authorities without asking questions. Pharmaca encourages people to take their health back into their own hands, an empowering concept.

STEPS FOR CONSCIOUS CONSUMERS

1. If you're starting a drug, investigate how to mitigate the potential side effects. Going on antibiotics? Maybe look into a probiotic you can take. Going on a cholesterol-lowering drug? Maybe supplement with COQ10. Going on a proton-pump inhibitor to treat heartburn? Perhaps supplement with B12.[89]

2. Re-evaluate the quality of the vitamins you take, because you might be doing more harm than good. The growing multivitamin industry is worth billions of dollars,[90] and 68 percent of Americans take them.[91] Unfortunately, many vitamins on the shelves of stores today contain heavy metals, unhealthy additives, and genetically modified ingredients. Labdoor, a consumer protection-oriented company, tested thirty-four of the most common American magnesium

---

89 "Are Your Medications Causing Nutrient Deficiency?" Harvard Health. Harvard University, August 2016.
90 "U.S. Revenue Vitamins & Supplements Manufacturing 2019." Statista. Statista, n.d.
91 "THE DIETARY SUPPLEMENT CONSUMER: 2015 CRN CONSUMER SURVEY ON DIETARY SUPPLEMENTS." CRN The Science Behind Supplements. CRN, n.d.

supplements and found that twenty-five of the supplements had levels of arsenic that surpassed the recommended limit of daily arsenic consumption, according to California's Proposition 65.[92] Proposition 65, enacted in 1986, protects the consumer by requiring companies to label their products with warnings if they contain unsafe levels of chemicals the State of California deems to be dangerous.[93] To avoid heavy metal contamination, research the brand you are considering and check to see how that brand tests its products for safety. Third-party testing is best.

3. To avoid genetically modified ingredients, the simplest solution is to buy only organic supplements. However, healthy supplements are sometimes not certified organic and can still be safe. So, if the supplement in question is not labeled organic, make sure to research beforehand. Research if you see the following corn-derived ingredients: vitamin C, ascorbic acid, whey protein and gelatin (cows sometimes eat corn), citrate minerals, citric acid, maltodextrin, corn starch[94], xanthan gum, and ascorbyl palmitate.[95] Delve deeper if you see the following

92 "Top 10 Magnesium Supplements." Labdoor. Labdoor, n.d.
93 "About Proposition 65." OEHHA. OEHHA, n.d.
94 Tompkins, Tiffany. "Supplement Law Makers Brace for Federal GMO Labeling Law." Compass Natural Marketing. Compass Natural Marketing, October 8, 2016.
95 Vien, Anya. "Shock Finding: Top Pharma-Brands of Vitamins Contain Aspartame, GMOs, and Hazardous Chemicals." Anya Vien, May 4, 2019.

soy-derived ingredients: vitamin E, soybean oil, and lecithin. Research further if you see the following genetically modified ingredients: B2, B12, beta carotene, and amino acids.[96]

4. Too many potentially problematic ingredients are sold in supplement aisles today, so research them individually before purchasing. Artificial coloring, which is potentially carcinogenic, should be avoided.[97] Magnesium silicate, which can be contaminated with asbestos, and inflammatory and immunosuppressive titanium dioxide are two other additives that should be skipped.[98]

96 Tompkins, Tiffany. "Supplement Law Makers Brace for Federal GMO Labeling Law." Compass Natural Marketing. Compass Natural Marketing, October 8, 2016.

97 "Summary of Studies on Food Dyes." Center for Science in the Public Interest. Center for Science in the Public Interest, n.d.

98 Yigzaw, Erika. "5 Dangerous Ingredients in Your Vitamins and Dietary Supplements: Achs.edu." American College of Healthcare Sciences. American College of Healthcare Sciences, December 2, 2016.

CHAPTER 6

BEFORE I PUT ON MY MAKEUP

———

The average woman applies twelve products with 168 different ingredients onto her body each day, while the average man applies six products with eighty-five different ingredients every day.[99] How many of those products are safe?

According to the Environmental Working Group, 12,500 of the 82,000 registered chemicals in personal care products may have carcinogens, endocrine disruptors, pesticides, and reproductive toxins. That means 15 percent of what you're applying to your cheeks, lips, torso, or limbs every day could

———

99   "Exposures Add up – Survey Results | Skin Deep® Cosmetics Database." EWG's Skin Deep. EWG, n.d.

be contributing to your toxic burden, interfering with fertility, energy, mood, and overall health.[100]

As mentioned previously, the FDA does not thoroughly screen each ingredient before products are released. In fact, the FDA website states, "With the exception of color additives and a few prohibited ingredients, a cosmetic manufacturer may use almost any raw material as a cosmetic ingredient and market the product without an approval from the FDA."[101] As of 2009, the European Union banned about 1,300 chemicals from cosmetics, Canada banned 600 as of 2018, and the United States has banned a mere thirty.[102]

Of course, you're not chugging your aftershave or nibbling on your eyeshadow. But you still may be absorbing toxic chemicals. In fact, skin is the largest organ of the body and has the capacity to absorb everything put on it. Examples include nicotine or birth control patches. People apply these patches to their skin, so that they can absorb substances directly into the bloodstream.

---

100 Ibid.

101 "Cosmetics Safety Q&A: Prohibited Ingredients." U.S. Food and Drug Administration. FDA, n.d.

102 "Why Beautycounter Bans More Ingredients Than The U.S." Beautycounter. Beautycounter, August 1, 2018.

HONEY AND VINEGAR

Do not despair. Hope exists. Meet Elise Graham Kennedy—founder of Honey and Vinegar non-toxic makeup line, blogger, entrepreneur, graphic designer, disability advocate, and fourth-generation Austinite. Gwyneth Paltrow, actress and founder of Goop, a natural health company, once proclaimed, "I'm in love with her," after Elise pitched her finance app, Olivia A.I., on a reality show called *Planet of the Apps*.

Elise and I crossed paths the way most Millennials meet these days—through the world wide web. Despite liking each other's photos and my admiring her from afar, we had actually never spoken. Elise and I found each other on Instagram, because we are both part of the online Lyme community. People from all over the world with Lyme disease, as well as people with other chronic illnesses and disabilities, connect on Instagram for friendship and support.

For Elise, Instagram is an extension of her blog, Honey and Vinegar, where she has documented her health journey, recipes, and even her experience on another television show, *The Price is Right*. In June of 2018, Elise launched her make-up line, also called Honey and Vinegar, which she created after doctors advised her that she had to switch to all-natural products after stem cell treatment for Lyme disease. As Honey and Vinegar's slogan says, Elise wants consumers to break up with their toxic make-up and switch to "clean, accessible, and affordable toxin-free" products.

She created a line of beautiful liquid lip colors: Blush Your Heart, Chestnut, Patty Gene, Sassafrass, Zoeybug, Smooth-talker, Optimist, Malibu, and 9 to 5. She also sells a body oil called Yellow Rose of Texas and two eyeliners, Onyx and Sienna. She tells me in an interview, "Personally, I started with liquid lips because that's what brought the most joy to my life. I wanted something that would stay on. With chronic illness, you have more things to worry about."

The name, Honey and Vinegar, originates from something her grandmother, whom Elise calls one of her best friends, always says—you can catch more flies with honey than you can with vinegar. She wants her make-up line to be a "love letter to women everywhere. An ode to the strong women in my life that came before me."

While Elise grew up in a health-conscious household in small-town Texas with a father who was a physician, she did not switch to all-natural living until someone at her stem cell clinic started educating her about the ingredients in American conventional make-up products. Elise says, "Once I knew, I couldn't go back to the products I had been using before."

Upon returning home, she began her "crunchy-living" initiation. She visited a local health food store, assuming the products there would be safer. However, when she opened her Environmental Working Group app, she was disappointed that these products rated lower than she'd expected. Later, online, she found some products that had higher safety standards, but she still was not pleased with them, telling me, "I felt like I was sacrificing the thing that I loved so much—feeling beautiful while being sick—that was really important to me. I felt like I had to make a huge sacrifice, which was kind of depressing for a while." Elise is all about petite joys and bringing fun into the everyday routine.

Because Elise was continuing to share her health journey online after stem cell treatment, her followers started asking her if she had any favorite natural make-up recommendations, and she did not. She says that people asking for recommendations was the turning point. Clearly a problem existed, and she was determined to provide a solution: "That's when I started making cosmetics as a, 'Who knows if anyone will actually buy this?' It was more of a passion project for myself, because I needed it so desperately. My first batch sold out within twenty-four hours, and that's when I really knew I was onto something."

When I ask her about her thoughts regarding disability and entrepreneurship, she tells me her attitude has changed over the years:

I used to have a pretty poor relationship with [disability]. Because I thought it was something that was kind of my deep dark secret, and I was less worthy because of it. For the longest time, I said, "This doesn't define me." We're all multi-faceted creatures. I'm not a fan of ableism and inspiration porn, where people with disabilities only exist to warm the hearts and minds of able-bodied people. I don't want people to view me as disability first, because there are so many things that we have to offer as a disability community in this world rather than "being inspirational." So when someone thinks of me, I don't want it to be, "Oh, she has a disability, she's a great disabled entrepreneur," because for some reason, they are trying to put you in a certain box.

When I ask her how she navigates illness and work, she tells me her attitude now is healthier:

They're not limitations, it's my new capacity. Before, I compared myself to normal people that would work eighty hours a week, even fifty hours a week in an office, and I realized I can't and don't operate on that, but I am still as worthy, and I have as much value to bring, even if I don't fit into that box.

Chuckling, she says:

There's this quote that goes, "Entrepreneurs are the only people willing to work ninety-hour weeks to avoid forty-hour weeks." It's so true. I think for people like us, it's really stressful to have someone over you setting expectations when the biggest part of all of this is the most unpredictable, and that's your health. You're expected to know this pattern and perform a certain way, and for me, that was very stressful for a very long time. "How am I going to do this? How am I going to force my way through this?" I work a lot, but to be able to take a break in the middle of the day for a nap or to take a walk is an absolute game changer.

Elise says that she lets customers help pick what's next for Honey and Vinegar:

For me, this is such an act of love—toward you guys, toward the community. This is all for you. I really hope Honey and Vinegar grows more, because I want to hire people like you and my other friends who have Lyme and various illnesses and disabilities, because I think that's the way modern work is going—remote, flexible.

Elise envisions additional offerings, such as blush, bronzer, highlighter, brow pencil, lip balm, lipstick, and body care products. She has big plans for the brand and is adamant about making sure the ingredients in her products are safe. Her lip products are highly rated on the EWG website. Honey and Vinegar is on its way to becoming a non-toxic personal care empire.

STEPS FOR CONSCIOUS CONSUMERS

1. Remember that what you apply to your skin <u>will</u> be absorbed. Health is not exclusively about what you ingest.

2. Elise says to avoid the following ingredients, often found in personal care products: BHA/BHT, carmine, DEA anything, coal tar (CI-anything), parabens, parfum, phthalates, triclosan, PEG compounds, and synthetic dyes.

3. Additional non-toxic cosmetic brands Elise loves include Lilly Lolo and Juice Beauty.

4. Elise recommends a book called *Boundaries* by Henry Cloud, which helped her figure out how to balance entrepreneurship with illness.

CHAPTER 7

PLASTIC, MOVE YOUR BAGS, LET ME CALL YOU A CAB

———

As you're reading this book, so engrossed, you might have a water bottle nearby. You might take a sip, thinking about how you can't wait to leave a five-star review of this book on Amazon and tell all your friends and family about it. However, if your water bottle is made of plastic, we need to talk. Yes, even if it's "BPA-free."

Bisphenol A (BPA) is not the only plastic you should worry about, and "BPA-free" is greenwashing. In fact, in 2011, Environmental Health Perspectives, a peer-reviewed journal, published a study showing that even BPA-free plastic items can leach endocrine-disrupting chemicals. The authors of the study, led by George Bittner, a professor of Biology at the University of Texas, collected 450 plastic items from well-known stores, chopped up the items, and submerged the samples in either salt water or alcohol. Seventy percent of the products leached estrogen-like chemicals into the water at room temperature.[103] As you probably know, estrogen is the key female sex hormone, but everybody needs it. After adding heat to the samples, over 95 percent of the plastic samples leached estrogen-like chemicals into the water.[104]

Why would heat result in more leaching? Cheryl Watson, a professor at the University of Texas, says that heating up plastics is even more detrimental: "When you heat things up, the molecules jiggle around faster and that makes them escape from one phase into another. So the plastic leaches its component chemicals out in the water much faster and more with heat applied to it."[105]

Bittner founded a company, PlastiPure, to develop plastic products that were safer and had no estrogen-disrupting

103 Hamilton, Jon. "Study: Most Plastics Leach Hormone-Like Chemicals." NPR. NPR, March 2, 2011.
104 Ibid.
105 Pawlowski, A. "Left Your Bottled Water in a Hot Car? Drink It with Caution, Some Experts Say." Today. July 06, 2018.

compounds.[106] Bittner reported that Eastman Chemical's "Colorless Tritan bottles or Tritan bottles of colors other than green often exhibited leaching of chemicals with EA [estrogenic activity]."[107] Eastman sued PlastiPure because of that study. Despite Eastman Chemical labelling its products "EA-free," meaning estrogenic activity-free, PlastiPure continued to assert that Eastman's plastic products were unsafe. Eastman won the lawsuit, and PlastiPure was forced to change their marketing approach.[108] However, the results of Bittner's study are still alarming.

Here's why consuming estrogen-like chemicals is risky. They are endocrine disruptors, meaning they "may interfere with the body's endocrine system and produce adverse developmental, reproductive, neurological, and immune effects in both humans and wildlife."[109] Endocrine disruptors can also cause infertility.[110] These man-made "hormones" confuse the

106 Hamilton, Jon. "Beyond BPA: Court Battle Reveals A Shift In Debate Over Plastic Safety." NPR. February 16, 2015.

107 Bittner, George D., Chun Z. Yang, and Matthew A. Stoner. "Estrogenic Chemicals Often Leach from BPA-free Plastic Products That Are Replacements for BPA-containing Polycarbonate Products." Environmental Health : A Global Access Science Source. May 28, 2014.

108 Hamilton, Jon. "Beyond BPA: Court Battle Reveals A Shift In Debate Over Plastic Safety." NPR. February 16, 2015.

109 "Endocrine Disruptors." National Institute of Environmental Health Sciences.

110 Marques-Pinto, André, and Davide Carvalho. "Human Infertility: Are Endocrine Disruptors to Blame? in: Endocrine Connections Volume 2 Issue 3 (2013)." Endocrine Connections. BioScientifica, October 19, 2018.

body significantly. In a study conducted by Health Canada, a government agency, researchers discovered that BPA in the body is linked to the production of more fat, meaning the ingestion of these plastics could lead to obesity.[111] Remember that study conducted by the Environmental Working Group about the chemicals found in the umbilical cords of babies? The EWG found BPA.

Plastic water bottles are not the only items containing endocrine-disrupting chemicals. Conventional register receipt paper contains BPA or BPA derivatives, such as Bisphenol S (BPS). Kudos to PCC Community Markets in Seattle for being the first grocery store in America to switch to completely non-toxic receipt paper called Alpha Free, which uses a thermal coating made of vitamin C.[112]

Additionally, plastic is a huge contributor to trash pollution. Each year, about eight million tons of plastic are thrown into the ocean. Marine animals then consume the plastic, are unable to digest it, and die.[113] The ocean is full of plastic. The Great Pacific Garbage Patch, a mass of plastic brought together by ocean currents, is now twice the size of Texas.

---

111 Bienkowski, Brian. "BPA May Prompt More Fat in the Human Body." Scientific American. May 29, 2015. .

112 "PCC Natural Markets Is First Grocer in the Nation to Offer New, Safer Customer Receipts at All of Its Locations." PCC Community Markets. 2014.

113 "Fact Sheet: Plastics in the Ocean." Earth Day Network. April 05, 2018.

Plastic pollution is not just found on the surface of the water. In 2019, researchers from the Monterey Bay Aquarium in California sampled deep ocean water and discovered the highest concentration of microplastics between 600 and 2,000 feet below the surface.[114] Microplastics are tiny pieces of plastic smaller than five millimeters in length.[115] Marine life and humans consume microplastics via the food chain and our water supply.

In a 2019 study released by the journal *Environmental Science and Technology*, researchers found that the average American consumes more than 50,000 pieces of microplastic every year and breathes in a bonus of 50,000. Those who use disposable plastic water bottles are projected to consume an additional 90,000 pieces of microplastics per year, but the researchers also found a high quantity in tap water. These numbers can vary depending on one's water source, but are concerning nonetheless.[116] Why would we produce plastics when we know all this? Well, for one, the plastic industry is worth

114 Weise, Elizabeth. "Turns out There's More Plastic Pollution in the Deep Ocean than the Great Pacific Garbage Patch." USA Today. June 09, 2019.

115 Shultz, David. "Americans Eat More than 50,000 Tiny Pieces of Plastic Every Year." Science. June 05, 2019.

116 Fox, Kieran D., Garth A. Covernton, Hailey L. Davies, John F. Dower, Francis Juanes, and Sarah E. Dudas. "Human Consumption of Microplastics." Environmental Science & Technology. June 5, 2019.

$1.1 trillion globally.[117] Our society revolves around plastic, and so many important items we use every day rely on the inexpensive material.

LIFEFACTORY

But wait. There's hope. Meet Pam Marcus—San Franciscan, neonatal physical therapist, public speaker, and co-founder of the glassware company, Lifefactory. I first discovered Lifefactory many years ago when I was attracted to the vibrant water bottle display at my local health food store. I bought my first twenty-two-ounce glass water bottle and have been hooked ever since. I bring my Lifefactory bottle with me wherever I go.

Pam started her company after being a pediatric physical therapist for many years. She was handling plastic baby bottles all day while she taught premature babies how to latch onto a bottle for the first time. When she started learning about the health risks associated with plastic, in addition to the negative impacts on the environment, she was shocked and immediately changed her lifestyle. She started refilling glass Snapple bottles as an alternative to plastic bottles in her own life. With her newfound awareness, every time she fed the babies at the hospital, she could not help but be upset.

---

117 "The Global Market For Plastic Products Will Be Worth $1.2 Trillion By 2020." Resource Recycling Inc.

She could feel the warm milk in the plastic bottles and would imagine the chemical compounds leaching, and in turn, the babies' delicate systems absorbing the chemicals. Distraught, she decided to act.

Pam reached out to architect and integrative designer, Daren Joy. Together, she and Daren decided to create glass bottles for infants. She tells me in an interview:

> What makes our story so authentic is that we both have the Lifefactory lifestyle, meaning we both care about our bodies, our families, and their bodies and their health. I was drinking out of old Snapple bottles, and he was eating raw foods and also only out of glass, drinking out of mason jars, way back before it was cool like now

They went to a trade show in September of 2007 in Las Vegas, Nevada, with 500 or so glass baby bottles for sale. She says:

> We were thinking, "I hope people like this." We were in the new designer showcase, and it was amazing. We had no idea. We were just bringing our idea out into the world. Both of us were not really planning on being entrepreneurs or having a business—we just thought it was a great idea, because this was our lifestyle and this is what

we believed in, and it was a great design. So, we took it there, and it just took off. We got our first twenty-six accounts there.

She continues:

> When we came back home, now that we had twenty-six companies that wanted to buy from us, it was like, "Oh, wow, now we have to make bottles." So it was all a self-taught thing, and we had to get things into gear, it had to kick my butt, I was like, "I have to learn how to source, I have to figure out how to get these bottles made, and the best way of doing it, and the most cost-effective way." I just did it.

Eventually, Pam and Daren expanded to water bottles for adults. Lifefactory exploded, and celebrities like Kirsten Dunst, Jessica Biel, Drew Barrymore, Reese Witherspoon, Julia Roberts, and Anne Hathaway have all been photographed holding Lifefactory bottles.

Pam realized the FDA was not going to protect consumers as much as people think:

> We really can't believe a lot of what we read or hear, and we need to investigate things even more.

Companies say, "Oh, we are now BPA-free." But people don't realize, they don't know the real story that BPS, BPF, all the other BPA derivatives are just as harmful or more harmful than BPA. That's where I think the public really gets tricked, and it's unfortunate. The majority of people don't know about that. They think, "Oh, it's BPA-free, oh, I don't have to worry, I can drink out of this plastic bottle."

Pam works two days a week as a physical therapist at Kaiser Hospital in San Francisco. While she previously taught babies how to bottle feed, she now focuses on positioning premature babies so they still feel like they are in the womb. She must control their environment, so they can grow in a healthy way. Pam tells me that she works with babies as small as the size of one's palm all the way up to forty-week-old babies. She tells me:

I am fortunate to work every Monday morning in a high-risk infant clinic, where a team of us sees the babies that have graduated from the nursery at six months, eighteen months, and three years. It's really a joy to see. I'd say most of them grow up to be perfectly normal and fine. Big happy babies. It's really nice.

She will often recognize the parents but not the babies, because they have transformed so dramatically. She wants these babies to thrive, and she understands that plastic is not healthy for anyone at any age.

Pam says that the experience of co-founding Lifefactory taught her a lot. Since she started the company at forty-five, she realized that people are multifaceted and do not have to dedicate their whole lives to just one profession or passion. She says, "If I can start a company this late in life, you can really do anything. If you have the passion and tenacity, the motivation to want to do something that means a lot to you, then you can do many things in life. You don't have just one talent." Pam believes that you can always change your mind and follow a new path, regardless of what you studied in school or thought you'd always be.

She tells me that working with babies definitely shapes the way she sees the world: "It really is an interesting point of view—if we can start earlier with so many of these things that are not good for us, we can really change not just our actions but actually physiologically, neurologically, make a big difference." Pam understands that early interventions can prevent great catastrophe in the future.

While Pam is talking about the health of premature babies, this analogy applies to our own health and that of the planet. Prevention is key, and the changes we make today can improve the future.

STEPS FOR CONSCIOUS CONSUMERS

1. Look around your kitchen. What are your food containers made of? Are you still storing your food in plastic? Start storing food in glass containers, drink out of glass, bring reusable utensils with glass or wooden straws everywhere you go. Reduce your plastic waste as much as possible.

2. Avoid buying canned items. Many canned food items and beverages are lined with an epoxy material that contains BPA. The EWG tested 252 canned foods in 2014 and found that 78 of them had BPA.[118] If you need to purchase canned food or drinks, make sure they are BPA-free.

3. If you work with receipts all day, suggest to management that your company switch to BPA-free receipt paper. An even better option would be to switch to electronic receipts only.

4. Go to the EWG Tap Water database online and look up the tap water quality in your area. The tap water in my area, Seattle, has seven contaminants above safety levels— all potentially carcinogenic. Tap water not only contains microplastics but can have chlorine, fluoride, heavy metals, pharmaceuticals, and chemicals that aren't good for us.

---

118 Geller, Samara. "BPA in Canned Food." EWG. June 3, 2013.

5. Invest in a reverse osmosis or other water filtration system. Unfortunately, a lot of water filters that people use regularly are not effective. Researchers at *Natural News*, a health news organization, tested twelve of the leading water filters on the market in America. The famous pitcher-style water filter, Brita, scored the lowest; in fact, water going through the Brita filter had a 33.9 percent increase in aluminum due to the materials being used. While the Brita filtered out a little lead, arsenic, and uranium, it left the majority of the contaminants in the water. The Big Berkey, a portable and relatively inexpensive brand, performed the best.[119]

6. Consider investing in a shower filter. While shower filters are not very effective as of 2019 (unless you install a full-house reverse osmosis water filtration system), some can at least filter out chlorine. When chlorine is added to the water supply, it mixes with other pollutants and forms unhealthy and potentially carcinogenic substances.[120] Chlorine is also not the best beauty treatment, drying out skin and hair.

7. Worried about BPA already in your body? Sweat it out. In a 2012 study published in the Journal of Environmental

---

119 "Independent Laboratory Testing Results of Popular Gravity Water Filters."
120 Sharp, Renee, and J. Paul Pestano. "Water Treatment Contaminants:" EWG. EWG, February 27, 2013.

and Public Health," researchers at the University of Alberta-Edmonton, found that participants would sometimes excrete BPA via sweat even when the chemical did not show up in their blood or urine. This observation suggests that sweating could be the best way to excrete BPA.[121] To learn more about the benefits of sauna therapy, check out Chapter 18.

121 Genuis, Stephen J., Sanjay Beesoon, Detlef Birkholz, and Rebecca A. Lobo. "Human Excretion of Bisphenol A: Blood, Urine, and Sweat (BUS) Study." Journal of Environmental and Public Health. Hindawi, 2012.

CHAPTER 8

WHAT YOUR GYNO DIDN'T DISCLOSE

———

I barely remember health class when I was fourteen. We watched a video of a woman in a white hospital gown giving birth, screaming in pain. It looked horribly traumatic. We definitely learned about sexually transmitted diseases (STDs), and we practiced putting condoms on bananas.

The most vivid memory I have of the class is taking home an electronic baby for the weekend. Electronic babies looked like dolls except they had sensors that would keep track of our parenting skills. Each of us would "feed" a baby, "burp" a baby, and put a baby "to sleep." Sometimes the baby would wake up in the middle of the night and "demand" to be rocked. If we did not hold the baby in the proper position, supporting its neck in a certain way with our hands, its neck would "break," and we'd get a failing grade. The electronic babies would keep track of our attentiveness, and the teacher would deduct points accordingly.

My peers and I quickly learned that duct-taping the neck of the electronic baby was best—no broken neck, no bad grade, win-win. (For the record, if anyone is considering me for a childcare position, I promise I will not use duct tape. Unless absolutely necessary.)

I cannot remember learning about women's health, the menstrual cycle, or the fact that a woman is only able to conceive for an allotted amount of days each month—six days, in fact.[122] All I do remember with clarity is being exhausted by my electronic baby's middle-of-the-night crying and the sound of duct tape being pulled off the roll.

THE PILL

When I was a teenager, some of my friends began using hormonal birth control pills to help their acne or period symptoms. Dr. Jolene Brighten, naturopathic physician and author of *Beyond the Pill, A 30-Day Program to Balance Your Hormones, Reclaim Your Body, and Reverse the Dangerous*

---

122 "How to Chart Your Cycle to Know When You Can Get Pregnant." WebMD. WebMD, n.d.

Side Effects of the Birth Control Pill, tells me in an interview that there are two types of oral contraceptives—the combination and the progestin-only:

> The combination pill contains both synthetic estrogen and progestin. And the progestin-only is exactly what it sounds like and does not have estrogen. How do birth control pills work? By suppressing the signals from your brain to your ovaries and altering the natural function of the reproductive system.

Dr. Brighten says in her book that 60 percent of women go on the pill for non-contraceptive reasons.[123] She advises women to find the root cause of their hormonal symptoms rather than using a medication to mask them temporarily. She shares with me:

> Okay, so you now know that the main mechanism of birth control is to shut down your natural hormones. With that in mind, how could the pill fix a hormone imbalance when it is preventing you from making hormones in the first place? The truth is, the pill will not fix your hormone

---

123 Brighten, Jolene. *Beyond the Pill: A 30-Day Program to Balance Your Hormones, Reclaim Your Body, and Reverse the Dangerous Side Effects of the Birth Control Pill.* Harper One, 2019.

imbalance, your periods, your acne, your PCOS [polycystic ovarian syndrome]. It will not fix your hormones. Every day in my naturopathic medical practice I hear from women who have been told that taking hormonal birth control is the *only* solution if she wants to "fix" her hormones. That's B.S. Here's the deal: the pill, patch, IUD, (fill in the blank), all suppress your hormones. These devices tell your brain to stop talking to your ovaries and tell your ovaries that they've been replaced. And while these methods can strong-arm your body into submission and make those symptoms be gone, they are a short-term (and short-sighted) solution to your hormone struggles. Those symptoms that you hate? That we all hate? They are your body giving you some serious data that can help you fix your own period and make your menstrual cycle work for you.

Many women experience side effects while taking the pill, which Dr. Brighten describes as follows:

- "Hormonal confusion: missing or irregular periods, light or heavy periods, short cycles, infertility, headaches.

- Digestive problems: leaky gut, gut dysbiosis, inflammatory bowel disease.

- Energy reduction: fatigue, adrenal and thyroid dysfunction.

- Skin issues: hair loss, dry skin.

- Mood disruption: depression, anxiety.

- Lady part disturbance: low libido (Oh, hell no!), vaginal dryness, chronic infection, pain with sex.

- Vitamin, mineral, and antioxidant depletion such as folate, B12, and magnesium."[124]

In addition to side effects, Dr. Brighten warns of significant long term risks of the hormonal birth control pill, including thyroid issues, blood clots, heart attack, autoimmune disease, and cancer of the breast, cervix, and liver.[125]

The hormonal birth control pill can also alter who you find attractive. In 2008, researchers at the University of Liverpool discovered that the male scent to which women were attracted varied depending upon the type of contraception they used. Women on the pill found men with immune systems similar to their own more attractive, while those not

---

124 Brighten, Jolene. *Beyond the Pill: A 30-Day Program to Balance Your Hormones, Reclaim Your Body, and Reverse the Dangerous Side Effects of the Birth Control Pill.* Harper One, 2019.
125 Ibid.

on hormonal birth control were more attracted to men with different immune systems. The researchers reasoned that women naturally find men with different immune systems attractive, because their offspring will have a better chance of fighting infection were they to reproduce. The immune system genes a woman can smell in a potential partner are called the major histocompatibility complex (MHC).[126]

The lead researcher, Dr. Craig Roberts, says, "Not only could MHC similarity in couples lead to fertility problems, it could also ultimately lead to the breakdown of relationships when women stop using the contraceptive pill, as odour perception plays a significant role in maintaining attraction to partners."[127] Having a better romantic life is a potential benefit of avoiding hormonal birth control.

University of New Mexico researchers in 2007 conducted another fascinating study about attraction and birth control. They went to strip clubs in the area and tracked the tips strippers earned. Strippers on the hormonal birth control pill made considerably less money than those who were not on the pill. Dancers at the most fertile time of the month made $70 on average per hour, while those who were on the pill

---

126 Association, Press. "Contraceptive Pill 'Can Lead Women to Choose Wrong Partner'." The Guardian. Guardian News and Media, August 13, 2008.

127 Ibid.

made $37.[128] Apparently, men subconsciously are aware of the biological impact of hormonal contraception.

Even discontinuing the pill can create additional complications. Dr. Brighten writes in her book that women who stop taking the pill in order to conceive often experience many symptoms she calls Post-Birth Control Syndrome, also known as PBCS. She writes:

> PBCS symptoms can range from hormonal irregularities, which may include loss of menstruation, infertility, pill-induced PCOS, and hypothyroidism, to gut dysfunction and autoimmune symptoms. PBCS generally occurs within the first four to six months after discontinuing the pill and, in my experience, doesn't just go away without taking the necessary steps . . .[129]

These "necessary steps" include modifying one's diet, lifestyle practices, and supplementation.

Are these risks and symptoms worth it? After all, the hormonal birth control pill is 91 percent effective in preventing

---

128 Popsci. "New Study: Fertile Strippers Make More Money." Popular Science. Popular Science, October 5, 2007.

129 Brighten, Jolene. *Beyond the Pill: A 30-Day Program to Balance Your Hormones, Reclaim Your Body, and Reverse the Dangerous Side Effects of the Birth Control Pill.* Harper One, 2019.

pregnancy with typical use.[130] What if you could have a similar or better success rate and avoid health risks and potential side effects?

THE DAYSY

Developed in Germany, the Daysy monitor is a fertility tracker that can tell a woman with over 99 percent accuracy when she can conceive.[131] The device is a sophisticated thermometer that connects with a smartphone app. A woman simply takes her basal body temperature under the tongue first thing in the morning with the Daysy. Because her temperature is slightly higher during ovulation, the device tells her if she can have unprotected sex that day or needs to use a barrier method, such as a condom, to avoid pregnancy. The Daysy analyzes the temperature using an algorithm that includes data from five million menstrual cycles and thirty years of research.

If a woman is fertile, the Daysy will show a red light. If she is not fertile, the Daysy will show a green light. If the Daysy is still learning her cycle, it will show a yellow light. The Daysy takes two to four cycles to fully know a woman's fertile

130 "How Effective Is Contraception at Preventing Pregnancy?" *NHS Choices*, NHS, 30 June 2017,
131 "Natural, Self-Determined Fertility Tracking with Daysy." Daysy. Daysy, March 12, 2019.

window, so at first, it will be more conservative and show more yellow days. As a woman uses it more, the fertile window will become smaller. And, if a woman is trying to get pregnant, she can know with much greater accuracy when she should have sex to conceive.

Meet Natalie Rechberg, Daysy's founder. Her father, Dr. Hubertus Rechberg, invented the BabyComp in 1986 in Germany. It was the first computer fertility monitor ever created. Working alongside a team of gynecologists, engineers, and software specialists, he invented it to help his wife avoid the negative effects of the hormonal birth control pill.

The BabyComp, which was initially the size of a grapefruit, became the LadyComp, and then transformed into the Pearly. The Pearly was no longer circular but oval-shaped and fit in the palm of one's hand. Each iteration was smaller and improved, but Natalie tells me in an interview that the previous technology hadn't been as desirable for the modern woman. The Pearly showed temperature values, but she wanted lights to tell her whether she was fertile that day. She also knew that the modern woman uses a smartphone for all sorts of daily tasks, so why not have a fertility monitor connect to it, too? It just made sense.

Natalie, a self-proclaimed country girl, is the oldest of four sisters and grew up in the beautiful Bavarian Alps near

Munich, Germany. Natalie tells me that her father didn't talk much about his work. She says:

> I was super surprised myself when I realized what my father was doing. We grew up with my father—my parents divorced when I was eight—and he would not talk about periods or anything like that. When I entered the company, then we really had to start talking about sex. We started talking about the female menstrual cycle. I had the feeling he didn't feel comfortable talking to his daughters about it.

Natalie thinks that talking about women's health topics might be becoming less stigmatized in Germany, in addition to Switzerland, where she lives now. She reflects:

> Women are starting to question what they have been told by medical authorities and do their own research. I feel there's a completely new movement. These women really question what's out there. We don't swallow anything anymore just because it is prescribed to us. We are gaining in strength in the sense that we are feeling comfortable with our bodies and being conscious of our bodies, and this makes up powerful females.

Women today also have access to many more online resources to learn about women's health and fertility.

Natalie went on the pill when she was a teenager because all her friends were doing so, and she did not feel comfortable talking about the topic with her father. She says, "I had severe side effects, which damaged my body." Then she tried the copper IUD. In 2008, when she started working at her father's company, she delved deep into the world of the Fertility Awareness Method (FAM), which is the basis for her father's technology. She then had her copper IUD removed and completely switched over to natural options. Now Natalie uses her Daysy daily. She tells me:

> I use the Daysy to track my cycle and know when I am in my fertile window and when I am not. We use protection during my fertile window. I used the Daysy to plan my two pregnancies. Timing is everything when it comes to conception. I told my husband that if we want to plan, now is the right time, and we fell pregnant both times in the first cycle trying.

Because Natalie planned her pregnancies with the Daysy, she knows the exact dates of conception, which made predicting her delivery dates even more precise. When I interviewed her, she was thirty-seven weeks into her pregnancy and still working at the customer service desk. Natalie is dedicated. She continues:

> I also use her [the Daysy] as a little scanner for health-related issues. So, if something is off in my cycle, I go and show it to my traditional Chinese medicine doctor (TCM), naturopath, or acupuncturist, who then tells me what's happening and how I could regulate it. I use her for planning my day-to-day. I know when in the menstrual cycle I should do what. So, I am planning important meetings during my fertile phase and not so much during my period. I schedule my calls depending on where I am in my cycle. I know when my body needs rest. I live my cycle. It gives me great awareness of myself.

Natalie and her husband, Collin, run the family business. Collin is "Mr. Finance" and supervises "overall operations, strategic and business planning, and quality control." Natalie, on the other hand, participates in "product development, strategic marketing, distribution, and sales." They share an office. She says, "I trust him, he trusts me, and we complete each other. We can rely on each other and that is very important when working together." She continues, "I love working for something which is now in the next generation. To let the vision of my father live on."

I ask her about the word Femtech, which is used to describe technology that helps women with their health. Other examples of Femtech include an artificial intelligence device that can screen women more effectively for cervical cancer than a standard Pap smear and an app for women with gestational diabetes to observe and communicate physiological changes to their doctors.[132] Natalie tells me that, surprisingly, Femtech is still a male-dominated field, and the word is relatively new. When she first went to medical trade shows with her father years ago, she was shocked by the lack of women. Women in Femtech are essential, because they are the ones who actually use the products.

Not only can the Daysy help plan and prevent pregnancy, but it can help women find root causes of their hormone-related health problems. For example, the Daysy assists in diagnosing hypothyroidism, polycystic ovarian syndrome (PCOS), cysts, and endometriosis, based upon temperature fluctuations. Natalie tells me that the Daysy is transformative and empowering for women because they can take their health into their own hands and understand their bodies, mood shifts, and symptoms in ways they never could before.

Natalie tells me that she recently interacted with a customer who used the Daysy to conceive her first child. While the

---

132 Das, Reenita. "Women's Healthcare Comes Out Of The Shadows: Femtech Shows The Way To Billion-Dollar Opportunities." Forbes. April 12, 2018.

woman had a positive pregnancy test, the gynecologist she saw told her that she wasn't pregnant, but rather, she had a cyst. The doctor gave her medication to make her menstruate again. The woman trusted her instincts, believing she was actually pregnant, and decided not to take the medication. Instead, she discussed her situation with a Daysy medical advisor and sought the expertise of another gynecologist. The second gynecologist confirmed a pregnancy. A picture of the baby proudly adorns Daysy's office wall.

STEPS FOR CONSCIOUS CONSUMERS

1. If you're a woman on the pill and want to stop taking it, check out Dr. Jolene's website, www.drbrighten.com, and her book. Her resources are also helpful for those who want to support their bodies, while staying on the pill to avoid pregnancy.

2. If you're looking for barrier methods, condoms are a healthy option. Check out the next chapter to learn about Sustain, a safe and sustainable condom company. Natalie is a fan of the Caya diaphragm with Contragel (a natural and non-toxic spermicide).

CHAPTER 9

LET'S GET IT ON

———

You now know which contraceptive options might be best for you or your partner. You're ready to dive deep into the healthiest and most sustainable sex life ever, right? Wait. Not yet.

Unfortunately, a lot of condoms and lubricants are not healthy or sustainable. Most condoms contain a chemical called nitrosamine, which is a carcinogen that is created when latex is heated and molded. The nitrosamines are liberated when a condom touches body fluids.[133]

Women are potentially more susceptible to absorbing chemicals than men. The vagina and vulva are mucous membranes, meaning "the vagina is capable of secreting and absorbing

———

133 Seppanen, Jahla. "Woah, Should You Be Using Organic Condoms?" Shape.

fluids at a higher rate than skin, as are some of the external portions of the vulva, including the clitoris, clitoral hood, labia minora, and urethra."[134] In fact, pharmaceutical companies are even designing drugs to be administered via the vagina rather than the mouth. In one study, women used estradiol, an artificial estrogen, either vaginally or orally. The results? The women who took the estrogen vaginally had ten times more estrogen in their blood than the oral participants.[135] The vagina is like a sponge.

In addition to the health risks from nitrosamines, natural latex condoms have other negative effects. Natural latex is made from the sap of the rubber tree. Demand for rubber is on the rise globally, and the vast majority of rubber is grown in Southeast Asia. To make rubber plantations, the area's biodiversity is destroyed, which means insect, bird, and bat populations are significantly reduced. Also, unfortunately, many rubber plantations use pesticides and herbicides, which make their way into waterways,[136] and child labor is often utilized. In fact, around 1.5 million Indonesian children between the ages of ten and seventeen work in the agricultural sector, which includes rubber.[137]

134 Nicole, Wendee. "A Question for Women's Health: Chemicals in Feminine Hygiene Products and Personal Lubricants." National Institute of Environmental Health Sciences. March 1, 2014.

135 Ibid.

136 Normile, Dennis. "The Tires on Your Car Threaten Asian Biodiversity." Science. December 10, 2017.

137 "Child Labour in Plantation." Child Labour in Plantation. April 22, 2010.

In addition to condoms, many people use products to enhance their sex lives. One of those products is lubricant. In 2009, researchers at Indiana University surveyed 2,453 women between the ages of eighteen to sixty-eight and found that over 65 percent of them had more pleasurable and comfortable sex because of lube.[138] In 2012, Indiana University surveyed over 1,700 men and women over age eighteen about their sex lives. Thirty percent of women surveyed said that they experienced pain during their last vaginal sexual encounter.[139] While there are many causes of painful sex, such as menopause, endometriosis, or vaginitis,[140] lube can make sex far less painful for many women.

Unsurprisingly, many lubricants have unhealthy ingredients you don't want near your private parts. Ingredients to avoid include glycerin, Nonoxynol-9, parabens,[141] and petrochemicals, such as propylene glycol, polyethylene glycol, and petroleum.[142] Glycerin can cause yeast infections, and

138 "Studies About Why Men And Women Use Lubricants During Sex." ScienceDaily. November 09, 2009.

139 Herbenick, Debby, Vanessa Schick, Stephanie A. Sanders, Michael Reece, and J. Dennis Fortenberry. "Pain Experienced During Vaginal and Anal Intercourse with Other-Sex Partners: Findings from a Nationally Representative Probability Study in the United States." The Journal of Sexual Medicine. February 04, 2015.

140 "Women's Health Care Physicians." ACOG.

141 Barnes, Zahra. "6 Lube Ingredients You Might Not Want to Put in Your Vagina." SELF. December 01, 2017.

142 Acciardo, Kelli. "4 Harmful Lube Ingredients You Should Avoid At All Costs." Prevention. June 10, 2019.

the spermicide, Nonoxynol-9, can kill both good and bad bacteria, disturbing the pH of the vagina and resulting in vaginitis. Parabens are preservatives that can potentially disrupt endocrine function and leave women at greater risk for breast cancer.[143] Finally, the petrochemical, propylene glycol, is used in anti-freeze and can cause vaginal irritation.[144] (Propylene glycol might sound familiar to you—it's used in the vaping liquid of e-cigarettes.)[145] Petrochemicals are also unsustainable because they are made from fossil fuels.[146]

SUSTAIN NATURAL

So, what do you do if you want to use condoms to protect yourself against STDs and prevent pregnancy? Do you need to start growing your own rubber trees? Never have sex? No, no need to worry.

Meet Meika Hollender, co-founder of Sustain Natural, a sexual wellness company that makes safe and sustainable sexual health products. Meika grew up immersed in the non-toxic product sector. In fact, her father, Jeffrey Hollender, is the

143 Barnes, Zahra. "6 Lube Ingredients You Might Not Want to Put in Your Vagina." SELF. December 01, 2017.

144 "What Are Petrochemicals? - The Honest Company Blog." The Honest Company Blog. January 03, 2019.

145 "Some e-Cigarette Ingredients Are Surprisingly More Toxic than Others." Medical Xpress. Medical Xpress, March 27, 2018.

146 "What Are Petrochemicals? - The Honest Company Blog." The Honest Company Blog. January 03, 2019.

founder of Seventh Generation, the well-known cleaning and home product company.

She tells me in an interview that she started interning for Seventh Generation in high school and worked there throughout college. In 2006, Meika and her father wrote a book, *Naturally Clean,* about non-toxic household products. She has always been passionate about ingredient transparency and sustainability. She tells me:

> It's funny, because I watched my dad start and run this incredible business that grew so much over time. I always thought to myself that I never wanted to be an entrepreneur because of how stressful it seemed. Looking back now, I remember how fulfilled my dad felt, and the moments of celebration versus the stress— or maybe I've trained myself to remember those things.

With the help of her father, Meika founded Sustain in 2013 because she saw a lack of safe sexual wellness products on the market. She says:

> I started Sustain because I thought women deserved better options when it came to their sexual wellness. Options that were not only healthy

for their bodies, but also options that celebrated their sexuality rather than shamed them for it.

Meika saw an unsettling dichotomy—women were oversexualized in mass media but often weren't given comprehensive sex education. Women felt embarrassed or judged if they took their sexual health into their own hands by buying condoms. Meika wanted Sustain to not only be a successful company, but a platform for activism as well.

In 2016, Meika launched a Sustain campaign called #GetOn-Top, in which 100,000 women pledged to practice safe sex and talk about it with their family and friends. Meika tells me:

> I launched this campaign because only 21 percent of single, sexually active women use condoms regularly, 41 percent of pregnancies are unplanned in the US, STDs are on the rise and disproportionally affect young women, and women are still slut-shamed for being sexual. Getting 100,000 women to pledge to practice safe sex, and even just admit they have and enjoy sex, was a powerful statement and a first step in moving these statistics in the right direction.

America is *uncomfortable* talking about sex. According to Guttmacher, a sexual health research institute, only twenty-four states and the District of Columbia require sex education in schools. Only thirteen states require that the education be medically correct, and only eighteen states and the District of Columbia require instruction about contraception.[147] Although research demonstrates that abstinence-only education programs lead to higher rates of teen pregnancy, the federal government funds these programs, allocating millions of dollars toward teaching ineffective practices.[148]

Sex attitudes and education are significantly different in other parts of the world. For example, the majority of Western and Northern European countries require sex education, often from a young age. In Scandinavian countries, such as Denmark, Norway, and Finland, sex education can begin in pre-school and continue through high school. Many countries in Europe avoid abstinence-only education and treat sex as a natural human act instead of vilifying it or simply not acknowledging it at all.[149]

The Netherlands has the second lowest teen pregnancy rate among developed countries, and unsurprisingly, sex education can occur there beginning at the age of four. In a

---

147 "Sex and HIV Education." Guttmacher Institute. June 03, 2019.
148 Stanger-Hall, Kathrin F., and David W. Hall. "Abstinence-Only Education and Teen Pregnancy Rates: Why We Need Comprehensive Sex Education in the U.S." PLOS ONE. October 14, 2011.
149 Suzdaltsev, Jules. "What Students in Europe Learn That Americans Don't." Vice. March 16, 2016.

viral video made by the production company Attn, a Dutch primary school teacher, Lenneke Braas, says:

> We're inclined to think if you raise the subject of sexual education, then you might encourage children to start having sex at an early age. However, this is not the case at all. In fact, their knowing about it actually makes them wait a little longer, as sex is no longer such a big deal.[150]

As of 2014, the abortion rate in the Netherlands was 8 for every 1,000 women,[151] compared with 14.6 in the United States.[152] The abortion rates in Denmark, Norway, and Finland are also lower than the United States—12 per 1,000 women in 2014, 12 per 1,000 women in 2015, and 8 per 1,000 women in 2015, respectively.[153] Meika admires the "syllabus that Planned Parenthood has tried to put forward in New York City around comprehensive sex education, focusing on birth control, pleasure, consent, sexuality, and communication."

Meika considers female pleasure a vital part of the sexual health conversation. In 2018, Meika published a book inspired

---

150 "We Can Learn a Lot from the Netherlands' Approach to Sex-ed." Facebook Watch.
151 "Abortion Worldwide 2017." Guttmacher Institute. 2018.
152 McCammon, Sarah. "U.S. Abortion Rate Falls To Lowest Level Since Roe v. Wade." NPR. January 17, 2017.
153 "Abortion Worldwide 2017." Guttmacher Institute. 2018.

by the campaign entitled *Get on Top: Of Your Pleasure, Sexuality & Wellness: A Vagina Revolution*. She informs me that only "32 percent of heterosexual women report having an orgasm every time they have sex, while about 92 percent of heterosexual men report the same."

Sustain sells menstrual products, including organic cotton tampons, pads, and liners, period cups, and period underwear. The company also sells massage oil, body wash, and lip and body balm. Meika recommends scrutinizing ingredient lists:

> Only buy products going inside your vagina that list every single ingredient. What most people don't know is that the FDA does not require tampon or condom brands to list their ingredients. Because of this secrecy, buying and using brands that *do* list their ingredients is critical.

She also recommends using the Environmental Working Group (EWG) database described in Chapter 3.

Meika attributes Sustain's success to its "hero product," the condom. The latex in Sustain condoms is sourced from the "only Fair Trade, FSC-certified rubber plantation in the world that makes latex for condoms." This plantation is located in India. Its lube is 96 percent organic, pH compatible, and free of petroleum, parabens, glycerin, and fragrance, and can safely be used with latex condoms.

She also credits Sustain's success to "authentic, action-oriented, and straightforward discussions and actions around female sexuality and women's reproductive health in the wake of the 2016 election."

Meika remembers when she first saw her products in a store. She was in a Pharmaca store in San Francisco: "I was with a friend whom I made take an incredible number of photos so that I could post our first #shelfie. It was totally surreal and exhilarating. I still get so lit up and surprised when I see us on a shelf!"

STEPS FOR CONSCIOUS CONSUMERS

1. Make sure your condoms are free of nitrosamines.

2. Check the ingredients of your lube and make sure they are non-toxic and vagina-friendly.

3. Oil-based lubes should never be used with condoms because the oil can break them down. Always make sure your lube can be used with condoms. Check the label.

4. Support political candidates who advocate for medi-cally-accurate sex education programs in schools and greater access to birth control and sexual health services.

5. Check out Meika's book, *Get on Top: Of Your Pleasure, Sexuality & Wellness: A Vagina Revolution,* for sex edu-cation you might have missed or forgotten from your school days.

CHAPTER 10

WHEN VAGINAS MET GLYPHOSATE

——

Did you know that the average, non-organic tampon or pad could contain pesticides, herbicides, carcinogens, allergens, endocrine-disruptors, and preservatives labeled under the ambiguous identifier "fragrance"?[154]

In fact, researchers at the University of La Plata in Argentina tested conventional tampons and pads and found that 85 percent of them contained glyphosate, a classified carcinogen,

---

154 Scranton, Alexandra. "Chem Fatale ." Women's Voices for the Earth: Creating a Toxic-Free Future. Women's Voices for the Earth, November 2013.

according to the World Health Organization.[155] See Chapters 3 and 13 to read more about glyphosate.

Tampons and pads are not regulated by the FDA, despite being categorized as medical devices. Feminine hygiene companies are not required to identify ingredients in their products,[156] so they can disclose or hide whatever they wish.

The average woman uses thousands of disposable products in her menstruating lifetime. HuffPost estimates that a woman uses 9,120 tampons during her life, which costs approximately $1,773.33, depending on the brand.[157] Women throw away billions of feminine hygiene products internationally.[158]

What should women do during that time of the month if they wish to be as sustainable and healthy as possible?

THINX

Meet Antonia Saint Dunbar, New Yorker, cello player, Kundalini yogi, co-founder and CEO of Antonia Saint NY shoes,

155 Graham, Karen. "Argentina Study: 85% of Tampons Contaminated with Glyphosate." Digital Journal. October 27, 2015.
156 Scranton, Alexandra. "Chem Fatale." Women's Voices for the Earth. November 2013.
157 Kane, Jessica. "This Is The Price Of Your Period." HuffPost. December 07, 2017.
158 Thorpe, JR. "It's Actually Possible To Use Pads & Tampons Sustainably - Here's How." Bustle. April 25, 2018.

and co-founder and former COO of Thinx. Thinx makes stain- and leak-resistant, anti-microbial, moisture-wicking, reusable period underwear. The underwear can be used in place of, or together with, a menstrual cup, pad, or tampon to prevent leaking.

In an interview, Antonia shares with me every menstruating female's worst nightmare—the horrifying leak. Antonia, then called Toni, was thirteen, and it was pajama day at school. She remembers being in biology class, wearing her favorite pink and white satin pajamas with a robe. It also happened to be the first day of her period, so she was wearing a pad. The running back of the school's football team leaned over and told her that she had sat in some ketchup. She continues:

> I was mortified. I got out of the room and went to the bathroom, found the very visible stain, and proceeded to wash out my pajamas. They became soaking wet, and I was just too embarrassed to go back to class. That was a traumatic moment for a thirteen-year-old girl to go through. That kind of thing is so preventable with a powerful back-up—the right product.

Twins Miki and Radha Agrawal first introduced to Antonia the idea of a stain-resistant underwear that could be used during a girl's period. The three friends were on a trip to

India for a wedding. Miki and Radha had come up with the idea years previously, and were also inspired by the TOMS shoes "One-for-One" model. TOMS donates a pair of shoes to a child in need with every purchase. The duo felt that such a solution would benefit girls in developing parts of the world.

The conversations among the three women in India inspired Antonia to quit her job as Director of Business Development at a recording studio in New York to focus on launching Thinx:

> I did what had to be done and quit my job in 2012, leaving behind years of contacts in an industry that had guided my career. I was taking the leap off this entrepreneurial edge with no parachute in sight. I lived off of my conviction that there was something huge to this dream, and I basically worked for nine months unpaid and with my husband Obed's support to get the Kickstarter video off the ground and to lead the development of the product.

Kickstarter is a crowd-funding site that allows entrepreneurs to introduce a product and raise money for its creation through donations from people passionate about the idea. Antonia and her co-founders thought this platform would be a good way to test the product, share it with the world, and see how people respond. She continues:

During this time, I was writing the patent and setting up all aspects of the business operations, from getting our LLC filed to getting a website set up. I was also working on finding and then managing our manufacturer, and I remember we had to change the cut and design of the product and its technology numerous times until we got it right, so that the performance would be where you need it, while also keeping you comfortable and dry.

Once the product was ready to market, Antonia and her co-founders launched the Kickstarter campaign, raising $65,000 in 2013. Shortly thereafter, Thinx won several awards, such as the Tribeca Disruptive Innovation Award, which focuses "on breakthroughs occurring at the intersection of technology and culture, where frequent clashes and resistance to change impede social progress and solutions for some of the world's most vexing problems."[159] Antonia also won free office space for six months via the Center for Social Innovation and Mass Challenge, a start-up incubator, where she and her co-founders were able to network with experienced executives and founders.

Thinx then won Crowdfund X, another crowdfunding challenge, which was funded by Nokia, a leading international

---

159 "What We Do." Disruptor Awards.

telecommunications company. At the Forbes Women's Summit, Miki and Rhoda spoke about the brand, and then met one of their future investors. Antonia adds, "These different connections were really great to have as our visibility was growing and as we sought out opportunities." As of 2019, Thinx has made it to the list of CNBC's Top 50 Disruptors, *Fast Company* magazine's "Top 50 Most Innovative Companies," *Entrepreneur* magazine's "Top 100 Most Brilliant Companies," and Number 37 on *Inc.* magazine's "5000 Fastest Growing Companies." Since Thinx's inception, the company has sold millions of pairs of underwear to women all over the world. Antonia shares that the secret to the success of her company has been the founders' passion to provide practical solutions for menstruating women in both the developing and developed worlds.

Thinx's mission from the beginning has not been merely focused on profit, but purpose, too. The company's founders were concerned that many menstruating girls in the developing world do not attend school because of their periods. In rural Uganda, for instance, researchers found that menstruating girls between the ages of twelve to seventeen missed 11 percent of school days due to "period poverty."[160] Period poverty refers to girls who cannot afford, or do not have access to, menstrual products. Antonia says:

---

160 Kampala, Dorah Egunyu in. "A Bleeding Shame: Why Is Menstruation Still Holding Girls Back?" The Guardian. May 28, 2014.

We wanted to create products of enduring value that performed really well, but also do good in the world. In the beginning, we had a partnership with Afripads, and we helped them expand their factories to produce and sell washable, reusable cloth pads to girls in Uganda.

Like the TOMS Shoes model, every time a customer bought a pair of Thinx underwear, Afripads would supply a woman in the developing world with a pack of seven pads. But Antonia clarifies:

We subsidized the cost of the pads, so they could still feel empowered to take care of themselves. We were not driving a welfare program, but an empowerment program. That first partnership with Afripads drove a lot of success in getting young women the products they needed, and the side effect was that it also generated good press and good will.

Missing school due to lack of period products is not just an issue in developing countries—inability to afford period products is also an issue in America and elsewhere. Nearly one in five girls misses school in the United States because she does not have access to period products.[161] In 2018,

---

161 "Join Our Mission to End Period Poverty | Always®." Join Our Mission to End Period Poverty | Always®.

137,700 girls missed school in the United Kingdom due to period poverty.[162] Menstrual hygiene products are sometimes too expensive or difficult to obtain for economically disadvantaged populations. To address this issue, Thinx launched a national grassroots campaign called United for Access, through which the company is petitioning the United States Department of Education to provide free menstrual products in all academic institutions across America.

Thinx's overall mission is dedicated to destigmatizing menstruation and creating solutions that empower and uplift their users. I learn from Antonia that the English word, taboo, comes from the Polynesian word, tapua, which actually *means* menstruation.[163] She says:

> In our society, women were not talking about the fact that they were bleeding and staining their underwear during their monthly cycles. So, we were just throwing away underwear. We also have companies that have ingrained in us the idea that we need all of these disposable products. They want to keep us hooked on disposable underwear, essentially, on the IV drip of plastic consumption

162 Elsworthy, Emma. "More than 137,700 Girls in UK Missed School in the Last Year Because They Couldn't Afford Sanitary Products." The Independent. March 7, 2018.

163 Grahn, Judy. "Chapter 1." Chapter 1: Blood, Bread, and Roses: How Menstruation Created the World.

of these wasteful, one-use feminine hygiene products. It is a huge toxic load for our planet. We have to do better for our earth.

Additionally, Thinx sells products through Speax, a urine incontinence underwear line. Every sale of Speax helps fund surgeries for fistula sufferers. Antonia explains:

> It's when a woman tears from childbirth, and she's too small—she could be pregnant at thirteen, fourteen-years-old, even younger—so her body isn't developed enough yet to have a baby. This can result in a tear, where she can leak urine and feces for the rest of her life, and she is done, ostracized forever, unable to work or care for herself properly. We have helped a lot of women already, but there is so much good that can be done. More companies need to be doing better and seeing a more holistic set of solutions that can be offered as a result of their business.

To remain balanced, Antonia says that waking up early and practicing yoga daily has helped her stay focused in the present moment, which sets the foundation of her day long before her two little girls wake up and she begins to read her many emails:

> You will easily have ten huge things going wrong in one day, and as an entrepreneur, you can have

daily reminders telling you that, "You should just quit. It's too hard. It'll be so much easier if you don't do this." So you have to ground yourself in what will make you rooted and strong. Purpose is also one of those things.

Everything is easier if you choose something that you're passionate about, and that you can make sacrifices for. The way to get through the challenges is that you cannot look too far ahead. Just focus on one thing at a time. That's how you don't get overwhelmed. There's also this deep place within, when you just know what you're doing is what you're meant to be doing, and if you feel that, then it gives you wings.

Millions of women worldwide wear Thinx products, and many people credit the company for launching the Femtech sector. Not only are women protecting themselves from glyphosate exposure and other undisclosed ingredients when utilizing reusable menstrual products, but they are dramatically reducing their waste. If the average woman has 451.3 menstrual cycles in her lifetime,[164] switching to reusable and safe options just makes sense.

---

164 Chavez-MacGregor, Mariana, Carla H. van Gils, Yvonne T. van der Schouw, Evelyn Monninkhof, Paulus A.H. van Noord, and Petra H.M. Peeters. "Lifetime Cumulative Number of Menstrual

STEPS FOR CONSCIOUS CONSUMERS

1. If you menstruate and are looking for sustainable period alternatives, consider investing in reusable period underwear, reusable organic cotton pads, or a menstrual cup.

2. If you are not ready to commit to reusable products, make sure the disposable products that you do use are organic. Remember, anything you put around or in your vagina will be absorbed. Choose your ingredients wisely. Remember that companies might not disclose what ingredients are put into conventional menstrual products.

Cycles and Serum Sex Hormone Levels in Postmenopausal Women." Breast Cancer Research and Treatment. PMC, March 2008.

CHAPTER 11

FAST FASHION, YOU'RE A LITTLE CLINGY

———

Keep reading. You are well on your way to becoming a healthier and more eco-friendly person. Perhaps by now you are buying more organic food to avoid glyphosate and other pesticides and insecticides. Maybe you're questioning the types of products you're applying to your body. Maybe you religiously drink out of a glass water bottle. Success.

However, there's an aspect of health and sustainability that many of us have never thought about. What am I talking about? Clothing. No, I'm not advocating for the construction of shirts made out of organic kale. I'm talking about sustainable fashion, as distinguished from "fast fashion," which is "inexpensive clothing produced rapidly by mass-market retailers in response to the latest trends."[165] Sustainable fashion takes into account people's health and the planet. This type of fashion considers the toxicity of the fabrics and dyes, how the workers who make the clothing are being treated, and the lifecycle of each garment (i.e., will it end up in a landfill or can it biodegrade?).

The estimated 2.4 trillion dollar fashion industry is quickly growing,[166] and some say this sector is the second most polluting industry in the world after oil.[167] While this ranking is often debated, clothing production is unquestionably polluting our world.

Shelbi Orme of the sustainability-focused YouTube channel, Shelbizleee, posted a helpful video about fabrics to purchase or avoid (see Chapter 19 to read an interview with her). She shares that the most unsustainable fabrics are polyester and synthetic fiber blends—polyester is literally made of a petroleum byproduct, ethylene.[168] When we wash polyester garments, they release microfibers, which are tiny pieces of fabric, into our water supply. These microfibers pollute our

165 "Fast Fashion: Definition of Fast Fashion by Lexico." Lexico Dictionaries | English. Lexico Dictionaries, n.d.
166 Amed, Imran, Achim Berg, Leonie Brantberg, and Saskia Hedrich. "The State of Fashion 2017." McKinsey & Company. December 2016.
167 Conca, James. "Making Climate Change Fashionable - The Garment Industry Takes On Global Warming." Forbes. December 03, 2015.
168 "Polyester." How Products Are Made.

oceans, marine life, and end up in our bodies, too. Furthermore, polyester and synthetic fibers will never biodegrade. Other unsustainable fabrics include leather (many chemicals) and non-organic cotton (many pesticides, a lot of land, and a lot of water).[169]

Mainstream clothing brands aren't protecting their workers and customers as much as they should. In 2012, Greenpeace International, an environmental non-profit organization, tested 141 garments from twenty-nine countries, including well-known brands such as Zara, Levi's, Gap, H&M, Victoria's Secret, Calvin Klein, Marks and Spencers, and Armani. Eighty-nine of the 141 garments contained Nonylphenol Ethoxylates (NPEs),[170] which cause reproductive and developmental harm in rodents and has been found in human breast milk, urine, and blood.[171] NPEs may be endocrine disruptors, harm aquatic life, irritate skin and eyes, and cause birth defects.[172]

All of the thirty-one screen-printed garments contained plastisol, a chemical used to apply logos and images onto clothing.[173] Plastisol is a phthalate, which in animal studies

---

169 Shelbizleee. YouTube. YouTube, April 24, 2019.

170 "Toxic Threads: The Big Fashion Stitch-Up." Greenpeace.

171 "Fact Sheet: Nonylphenols and Nonylphenol Ethoxylates." EPA. November 02, 2016.

172 "Get the Facts: NPEs (Nonylphenol Ethoxylates)." Safer Chemicals, Healthy Families. Safer Chemicals, Healthy Families, n.d.

173 "Toxic Threads: The Big Fashion Stitch-Up." Greenpeace.

has been shown to harm the liver, kidneys, and reproductive system.[174] Two of the garments tested from Zara contained azo dye,[175] a carcinogen banned in Europe.[176] The United States has considered an azo dye ban but has not yet implemented one.[177]

While these findings should concern the average consumer, the fact that workers have to be in close proximity to these carcinogenic substances is unacceptable. In Cambodia, which has many fast fashion factories that produce clothing for companies such as Gap and Adidas, thousands of workers have fallen ill. Rebecca Moss writes in *How We Get to Next* magazine, "Day after day, as harmful chemical dust and fumes are inhaled, ingested, and come in contact with skin, workers can experience immediate and long-lasting effects ranging from skin irritation and allergic reactions, to lung or organ damage."[178] Workers have also experienced mass fainting incidents, which may be related to toxic chemical exposure. In Cambodia, approximately forty or so dangerous chemicals are used to manufacture clothing and shoes.[179]

---

174 "Phthalates and DEHP." Healthcare Without Harm. Healthcare Without Harm, n.d.
175 "Toxic Threads: The Big Fashion Stitch-Up." Greenpeace.
176 "Are Your Clothes Poisoning You?" The Peahen. February 5, 2015.
177 "Complimentary White Paper." QIMA.
178 Moss, Rebecca. "Mass Fainting and Clothing Chemicals." How We Get To Next. September 29, 2016.
179 Ibid.

AMOUR VERT

Now, before you decide to join a nudist colony, meet Linda Balti, founder of Amour Vert (pronounced AH-moore VER) sustainable clothing brand. Linda was with her then boy-friend, now husband, Christoph Frehsee, on vacation in Peru in 2009 when she picked up a discarded copy of *Newsweek* magazine. She read the featured article about the environmental implications of the fashion industry and was absolutely shocked.

Linda grew up in Paris, and while she loved fashion, she had never really considered the environmental or ethical implications of the clothing she wore. That moment, reading the article in Peru, was monumental for her. She didn't know that she'd start a brand, but she knew there was no turning back. She tells me in an interview:

> I made the commitment from that moment to only buy either secondhand clothes or sustainable clothes. Same with food. I am not going to eat anything that is not going to be organic from now on. That was a huge change.

After ten months of traveling, Linda and Christoph returned home to Palo Alto, California, where Christoph was about to begin his MBA program at Stanford University. Before traveling with Christoph, she had been a computer engineer at a large defense company in France, helping design simulations for fighter jets and helicopters. While she had been successful in her previous profession, she did not feel passionate about it and knew she wanted to do something else. Linda decided to investigate sustainable brands because of the vow she had made to herself in Peru. The problem? Few affordable sustainable brands existed at the time.

As Linda continued to research sustainable fashion, she considered the idea of manufacturing dyes, textiles, and finishings for other brands. However, she quickly realized that most big brands had no interest in changing the operations of their supply chains. They were more focused on making money than sustainability. So Linda started designing sustainable textiles for her own brand, experimenting with various fabrics such as cotton, silk, and modal. Modal is a textile often sourced from birch trees that can be mixed with other fabrics to create a blend. She tells me, "Back in 2009, it was different, and I started understanding that the change would probably come from brands that were built on that sustainable DNA—where really, sustainability would be part of the brand."

At first, Linda decided to sell to retailers with the ultimate goal of transitioning to a direct-to-consumer platform. Going wholesale made more financial sense because the retailers commit to a specific quantity, meaning she didn't have to make extra garments that would never be worn. She sold to big-name department stores, such as Nordstrom, Neiman Marcus, Bloomingdales, and 500 to 600 specialty stores in America. She named her brand Amour Vert, which means "green love" in French. She says:

> We learned a ton about what was working—price points, the materials people were liking or not, fit, all of that, and trying to gather as much feedback as we could. Once we had all the data we needed, we focused on going direct to consumers, so e-commerce and retail. First opening a web store and then opening physical locations where people could experience the product and hear the story and try the garments.

Next, Amour Vert started a partnership with American Forests, a non-profit organization. With every t-shirt purchased from Amour Vert, American Forests plants a tree in the United States. Over 200,000 trees have been planted thus far. This partnership garnered the attention of celebrities. Hailey Bieber, actress and model, and Kourtney Kardashian, reality show star, posted about the partnership in honor of Earth Day. In 2012, the retail chain H&M released "Conscious Collection," a sustainable clothing line. This line helped Amour Vert make more sales because customers would search for sustainable clothing brands online and find Amour Vert's website.

Linda originally founded the company in her living room, then moved into an office two years later. Linda also brought in her husband, Christoph, after he graduated from Stanford in 2012. He became the CEO and took care of finances, accounting, and administration, while she focused on product innovation, design, supply chain, branding, and marketing.

In 2015, Linda opened the first Amour Vert store in San Francisco, then opened seven more stores across the United States. More celebrities started jumping aboard the Amour Vert train, which Linda attributes to helping her brand grow. Amour Vert partnered with Goop, actress Gwyneth Paltrow's lifestyle brand, and then partnered with Preserve, actress Blake Lively's brand. By the time she exited the company in 2018, there were seventy employees.

Linda is proud that her brand has helped spread awareness regarding sustainable fashion and the environmental repercussions of fast fashion. She says:

> People don't know these crazy facts, so if you don't know, you just buy whatever you find or whatever you can afford or whatever you like. That's one of the reasons why raising awareness and having that information more public helps people make better choices.

Linda shares that cotton production is responsible for 18 percent of the world's pesticide use, and 90 percent of the cotton in America is genetically modified. Only 10 percent is organic cotton. She continues:

> Not only are we polluting our soils, but we are also polluting crops, where those insecticides and those pesticides are living in the fiber, so it's what you are wearing. People have psoriasis, eczema, all those skin issues, it's totally crazy. Same with water usage.

She shares that farmers use over 5,000 gallons of water to grow 2.2 pounds of cotton, which only makes a pair of jeans and a t-shirt.[180] Linda tells me that she prefers incorporating hemp, which uses less water:

> We do quite a bit of that at Amour Vert—mixing linen and hemp and organic cotton. People still want the comfort—they want the softness, and I understand that. People don't want to run around in a potato sack. There are things that are possible today with finishings and everything to still get the performance, the softness, but lower the impact.

180 Thompson, Clay. "Ask Clay: Does Arizona Cotton Really Use That Much Water?" Azcentral. June 27, 2016.

Linda is also concerned about the treatment of garment workers. She says one in six people in the world works in the fashion industry, and many are extremely underpaid, as is the case in Bangladesh, where a lot of fast fashion factories operate. According to Oxfam, a poverty-fighting non-profit organization, nine in ten garment workers in Bangladesh cannot afford to feed their families.[181] She continues:

> When you know, you can't *un-know* all these facts, you can't ignore that and just go and buy whatever you find or you like, and that's really what happened to me. I think raising awareness is probably the most important thing because, again, if you don't know, you don't know. Once you know, then you are equipped to make better choices.

Linda emphasizes that she always wanted Amour Vert to grow quickly because she wanted to change the fashion industry. She says, "If you build a business that has one million dollars in revenue per year, it's great, it's a nice business, but you're not going to have a big impact because you're not going to change the game."

---

181 Haque, Moinul. "Poor Wages Force Bangladesh RMG Workers to Skip Meals." New Age | The Most Popular Outspoken English Daily in Bangladesh. February 27, 2019.

Linda is not a fan of trends and wanted to create timeless, non-disposable pieces. She says:

> We wanted to produce a product that she would keep for a very long time, which is the opposite philosophy of what fast fashion is about, where they produce a garment that is going to probably tear in six months. But that's the goal. They want you to come back and buy more. We also want you to come back and buy more, but we want you to keep what you have for a long time.

Amour Vert sells chic basics for women—casual, dressy, and workwear. Linda tells me that her favorite piece is a striped three-quarter length shirt called Francoise, which is named after Francoise Hardy, a famous French singer and fashion icon from the 1960s. Linda wears the top at least twice a week. Amour Vert also has a baby line and just released a men's line.

I ask her if she remembers when she saw someone wearing an Amour Vert piece for the first time. She tells me yes—it was like spotting a celebrity: "It was at the airport in the security line. That was crazy. I was with my husband, I was like, 'Oh my god, look, she is wearing one of our t-shirts.' My husband was like, 'Oh my god, I have to talk to her.' I was too shy." She says she was excited but didn't want to invade the woman's privacy. Her husband wasn't so shy: "We were queuing to get a coffee, and my husband went to her and was like, 'Your t-shirt is really cool.' She was like, 'Yes, it's Amour Vert.' He didn't say who he was! She was looking at him, like, 'Who are you?' That first time was emotional."

STEPS FOR CONSCIOUS CONSUMERS

1. Go thrifting. Buying vintage and used clothing is a great way to reduce your environmental impact. Make sure to wash your "new" clothing in a chemical-free laundry detergent, vinegar, and even Borax before wearing to remove all odors and mold.

2. Check out websites like Poshmark and Threadup, which are online marketplaces for used and vintage clothing. OfferUp and Letgo are also good websites for buying secondhand clothing from people locally. You can search for specific items or brands, which makes the process less overwhelming.

3. Download the free app Good On You. You can search by brand to see the ethics and sustainability practices of companies.

4. Trade clothing with your friends or attend a clothing swap in your area.

5. Learn how to sew, or meet your local tailor. If your clothing needs repairs, fix it or go to a tailor instead of just discarding the item.

6. Think about the materials in the garment you are considering. Avoid polyester and synthetic blends and opt for sustainable fabrics, including linen (made from flax plants), hemp, tencel (made of wood pulp), and organic cotton.[182]

7. If you already own polyester clothing, you can minimize your pollution when washing these fibers, according to the Plastic Pollution Coalition:

    a. Consider purchasing a Guppyfriend, which is a bag that holds your polyester clothing during the wash cycle.[183] The Guppyfriend website states, "Compared to washing without the Guppyfriend Washing Bag, 86 percent fewer fibers shed from synthetic textiles."[184] Also, the athletic clothing brand, Girlfriend, sells a microfiber filter, which you can attach to your washing machine.

---

182 Shelbizleee. YouTube. YouTube, April 24, 2019.
183 "15 Ways to Stop Microfiber Pollution Now." Plastic Pollution Coalition. March 02, 2017.
184 "What Material Is the Guppyfriend Made of and Is It Recycable?" GUPPYFRIEND.

b. When washing, use a cold setting. Cleaning your clothes in hot water will make them shed more.

c. Fill the machine when washing your polyester or synthetic clothing. More clothing creates less friction, which means less shedding.

d. Use liquid laundry soap rather than powder. Powder scrapes the clothing more, which means more microfiber shedding.[185]

---

185 "15 Ways to Stop Microfiber Pollution Now." Plastic Pollution Coalition. March 02, 2017. June 10, 2019.

CHAPTER 12

GET GROUNDED

———

What is grounding? It's a punishment inflicted by parents upon their angsty teenagers, right? Actually, no. Grounding, also known as earthing, is a health practice in which one places his or her bare feet onto the ground.

A quick refresher—the earth is negatively charged, and ideally we should be neutrally charged.[186] Our ancestors did not live in skyscrapers, hold electronic devices to their heads, or travel by plane. People slept on the ground, walked around barefoot, and foraged for food. It was really the crunchy golden age. I probably should have lived 40,000 years ago with the Neanderthals—I am 3.1 percent Neanderthal, after all. Thanks, 23&Me. But I digress.

———

186 "Does Grounding Really Work?" Bulletproof. December 12, 2017.

Because humans are no longer connected to the earth in the same way, we become positively charged and struggle to revert back to neutral.[187] Plus, we wear this fancy technology called the "shoe." While I appreciate the protection and warmth that shoes provide, our feet are constantly insulated with thick rubber and other materials. Unless you insist on walking to work barefoot (good luck with that), your body never gets the opportunity to connect with the earth's energy, return to neutral as nature intended, and normalize your Circadian rhythms.

In 2004, researchers gathered twelve participants who complained of insomnia, stress, and chronic pain. For six weeks, the researchers had participants sleep on grounded mats,[188] a type of earthing technology that allows participants to remain grounded while inside a building. The mat "plugs into the grounding wire port of a normal 3-prong outlet or a grounding rod .... The earth's natural electrons flow right up through the ground wire and onto the mat, even if you're in a high rise."[189] The researchers tested participants' cortisol levels, the main stress hormone of the body, before and after

187 Ibid.
188 Ghaly, Maurice, and Dale Teplitz. "The Biologic Effects of Grounding the Human Body during Sleep as Measured by Cortisol Levels and Subjective Reporting of Sleep, Pain, and Stress." Journal of Alternative and Complementary Medicine (New York, N.Y.). October 2004.
189 Wells, Katie. "Earthing & Grounding: Legit or Hype? (How to & When Not To)." Wellness Mama. January 23, 2019.

the six weeks. The results? Their cortisol levels improved, demonstrating stress reduction. The subjects also reported that their symptoms had either improved significantly or resolved entirely.[190] This study illustrates that grounding affects our physical and emotional wellbeing. We are intimately connected to the planet.

In 2015, thirty-two young and healthy men participated in a study demonstrating that grounding has the potential to reduce recovery time after exercise. Researchers split the group in half and had them exercise strenuously. Sixteen men were grounded post-exercise by standing on a grounding pad, while the placebo group was not grounded. Then, the researchers measured each person's Creatine Kinase (CK) blood values, an enzyme value that increases with high intensity activity.[191] Researchers found that on day two out of four, the CK levels of the placebo group rose, while the levels of the grounded group did not rise significantly.[192] Imagine if the participants had exercised while grounding. The body could recover even faster.

---

190 Ghaly, Maurice, and Dale Teplitz. "The Biologic Effects of Grounding the Human Body during Sleep as Measured by Cortisol Levels and Subjective Reporting of Sleep, Pain, and Stress." Journal of Alternative and Complementary Medicine (New York, N.Y.). October 2004.

191 "Creatine Kinase (Blood)." Creatine Kinase (Blood) - Health Encyclopedia - University of Rochester Medical Center.

192 Brown, Richard, Gaétan Chevalier, and Michael Hill. "Grounding after Moderate Eccentric Contractions Reduces Muscle Damage." Open Access Journal of Sports Medicine. September 21, 2015.

Health enthusiast and entrepreneur Dave Asprey, the founder of Bulletproof coffee, describes grounding in his 2017 book, *Head Strong: The Bulletproof Plan to Activate Untapped Brain Energy to Work Smarter and Think Faster in Just Two Weeks:* "One thing I have done for nearly a decade to increase my body's negative charge is called earthing, or soaking up negative charge from the ground. This can help your body build more EZ water." EZ water, also called exclusion zone water, was discovered by University of Washington bioengineering professor, Gerald Pollock. EZ water is the fourth phase of water after solid, liquid, and gas,[193] and is created in our cells. EZ water is essential to mitochondrial functioning and behaves like an antioxidant.[194]

Asprey continues, "Flying is a mode of travel that tends to reduce your negative charge, which lowers your amount of EZ water and causes inflammation. This is one reason you experience jet lag." He describes how he would feel less sick from flying if he grounded post-flight. He continues, "At the time, I had no idea why this worked, but now I know that I was soaking up negative charge from the earth and helping

193 Asprey, Dave. *Head Strong: The Bulletproof Plan to Activate Untapped Brain Energy to Work Smarter and Think Faster-in Just Two Weeks.* Harper Wave, 2017.

194 "What Is EZ Water and Why Do I Have to Get Naked In the Sun to Make It?" Bulletproof. February 18, 2019.

my body build EZ water. I noticed a marked difference when I practiced yoga barefoot on the ground."[195]

EARTH RUNNERS

Now all that sounds great, but what if we live away from nature? Meet Mike Dally, founder of Earth Runners, an earthing sandal company. Earth Runners sandals have grounded conductive laces, meaning they are lined with stainless steel thread that is connected to a copper plug. This copper plug, located on the bottom of the sandal, touches the earth and allows wearers to ground when walking on bare earth—grass, dirt, gravel, or rock. Earth Runners sells a variety of unisex minimalist grounded sandals for both adults and children. Customers can even buy kits to transform their existing shoes into grounding ones by installing a copper rivet.

After graduating with a degree in mechanical engineering from San Diego State University in 2010, Mike felt burnt out. He did not like the sedentary and conventional atmosphere of college, and he did not want a standard 9 to 5 job where his creativity could not be utilized. He took some time off, seeking an alternative, more meaningful route. Mike knew

---

195 Asprey, Dave. *Head Strong: The Bulletproof Plan to Activate Untapped Brain Energy to Work Smarter and Think Faster-in Just Two Weeks.* Harper Wave, 2017.

he ultimately wanted to spend time in a workshop, tinkering around. Mike, an avid runner, also discovered he liked the primal experience of running in sandals better than restrictive athletic shoes. So, he decided to design a minimalist sandal with grounding capabilities. Earth Runners was born.

In an interview, Mike tells me about his first Earth Runners prototype. He visited a local shoe supplier in San Francisco, and he felt like a kid in a candy shop:

> So, I just went and bought some stuff and started making sandals. I am a believer that you have to have a lot of bad ideas before you have a good idea, so you basically have to persist on prototype after prototype after prototype and try everything. It's almost like natural selection and evolution, you just have to go down every avenue. That's where it started. That's what I have continued to do.

Knowing that other minimalist sandals already existed, Mike's invention of the conductive laces was thrilling for him. After thinking of the lace technology, he felt "the tingles and this-could-be- big-type of feeling."

When I ask him about what makes a healthy foot, he shares his experience of gardening, comparing a foot to the roots of a plant. A plant growing in the freedom of a fabric pot is going to behave differently than one restricted to a plastic pot, which restricts the roots from breathing and integrating with the earth. He says:

> Look at baby feet when they are born. It's almost like a triangle—the heel being a point and then you got two other points—the big toe and the pinky toe, and it kind of angles out, but for us, it's looking like an oval, where the toes kind of arch in and compress into pointed toe footwear.

Mike then describes the effect of soccer on his feet: "In soccer, you're wearing super tight footwear and running—like extreme activity. So my feet went through a lot of deformation in that period. I feel like I have been reversing that damage for the past ten years." Mike's sentiment about babies echoes what many podiatrists and researchers conclude—that shoes are unhealthy for baby feet development.[196]

Mike continues: "So many bones, ligaments, and muscles have evolved for millennia mostly barefoot, and when you throw unnatural technology on it, you restrict the inherent natural movements, messing with many of the mechanics upstream in the body."

---

196 Murphy, Sam. "Why Barefoot Is Best for Children." The Guardian. August 09, 2010.

Mike believes that people have become disconnected from the earth and has dedicated his company to helping people "rewild" their lifestyle, a concept that means living in a more ancestral and natural way in alignment with our biology. Mike elaborates:

> Preserving healthy feet or rewilding domesticated feet can be achieved with natural footwear that aligns with the inherent functionality of the foot. Biomechanically sound, allowing for electrical connections, and not using synthetic, toxic materials and suffocating the foot. Those are all rewilding concepts—return biology to its more natural state. I feel like it's my answer to my happiness. The more I can rewild and be free, the happier I'll be.

Mike practices what he preaches. He spends time in nature as much as possible, running in the forest to find peace of mind and seek answers. He runs three to six times a week, from two to fifteen miles at a time.

When I ask him if he has any advice for aspiring entrepreneurs who aren't sure what they want to do, he recommends that they focus on what kind of customers they hope to attract:

> You're going to be interacting with and exchanging energy with your clientele on the regular, so

you want to reverse engineer who you do business with. So, if I want to connect with rewilding, empathetic, fitness-oriented, mindful people, it would be smart to choose something like Earth Runners. Creating a product for the type of people you want to connect with will make all the difference in the passion and satisfaction of your business ventures.

Mike now has six employees and spends his days in the unconventional, non-sedentary workplace he had always dreamt about. As they make the shoes, Mike and his employees listen to podcasts. A Tarzan swing hangs in the center of the shop, where workers can swing from ring to ring. He tells me, "It has been cool to have people come in, and they're like, 'Dude, it's so cool to work here!' I lose perspective on that a bit because it's all I've ever known, but I'm like, 'I guess it is pretty cool.'"

When Mike did not know what to do, he took a breather, went into nature, and found his path—a path he'd run while grounding on, obviously.

STEPS FOR CONSCIOUS CONSUMERS

1. Get outside more.

2. After an airplane flight, put your bare feet on the earth.

CHAPTER 13

WE FOUND LOVE IN AN ORGANIC PLACE

————

Organic food. Some say it's overpriced or even a scam, while others say it's a non-negotiable. According to Ecovia, a market research organization, the growing organic food market is estimated to be worth $97 billion worldwide. The United States has the biggest organic food market globally—worth an estimated $40 billion. Germany has the second biggest market, followed by France and China. In 2017 alone, organic farmland increased by 20 percent globally—over 269,000 square miles of farmland was organic.[197] That's more than the size of Texas.[198]

———

197 "GLOBAL ORGANIC AREA REACHES ANOTHER ALL-TIME HIGH." IFOAM Organics International. IFOAM Organics International, February 13, 2019.
198 "List of US States By Size." List of US States By Size, In Square Miles, n.d.

The demand for organic food is increasing. According to the USDA, "Organic sales account for over 4 percent of total US food sales."[199] In America, three out of four conventional grocery stores have organic options.[200] In a 2014 Gallup survey, based upon a sampling of over 1,000 people, 45 percent of people incorporate organic food into their diets. Younger people, those in urban areas, and those who make more money are more likely to incorporate organic food into their diets.[201]

Despite the rise of organic food globally, the majority of worldwide farmland is conventional, meaning pesticides and fertilizers are used. In fact, only 1 percent of farmland in America is organic.[202]

BARRIERS TO BUYING ORGANIC FOOD

PRICE POINT

In America, organic food is generally pricier, but not always. In 2015, Consumer Reports compared the prices of 100 different food items, both conventional and organic, from eight

---

199 "Organic Market Overview." USDA ERS - Organic Market Overview.
200 Ibid.
201 Gallup, Inc. "Forty-Five Percent of Americans Seek Out Organic Foods." Gallup.com.
202 "Only 1 Percent of US Farmland Is Certified Organic. Why Aren't More Farmers Making the Switch?" Genetic Literacy Project. January 05, 2019.

well-known American grocery stores. On average, organic food was 47 percent more expensive—however, in some cases, the organic product was the same or lower priced than the conventional. For example, conventional and organic carrots were the same price at Safeway, and conventional maple syrup was more expensive than the organic equivalent at Price Chopper.[203]

In 2013, according to the US Department of Agriculture, Americans spent a lower percentage of their income on food than previously. Americans spent an average of 9.9 percent of their income on food—these rates have been decreasing since the 1960s. This decrease is attributed to lower food prices.[204]

NOT BELIEVING IT'S HEALTHIER

Additionally, among those who can afford organic food, many do not see the point of spending more on a food item that tastes and looks the same. However, researchers have connected eating non-organic food to a higher risk of health issues. In a 2018 study in JAMA Internal Medicine, a peer-reviewed journal published by the American Medical Association, French researchers surveyed over 68,000 people and found that those who ate organic food were 25 percent less

---

203 "Cost of Organic Food - Consumer Reports." Cost of Organic Food - Consumer Reports.
204 Barclay, Eliza. "Your Grandparents Spent More Of Their Money On Food Than You Do." NPR. March 02, 2015.

likely to have cancer.[205] Those who eat organic food are probably more likely to adopt other health-conscious practices, which could further reduce their risk of cancer.

Furthermore, in a study published by the British Journal of Nutrition in 2014, researchers found that conventional food had 20 percent to 40 percent lower levels of antioxidants than the organic versions. Antioxidants are cancer-fighting and anti-aging compounds.[206]

The impact of "Blue Zones" further suggests that eating organic food is healthier. The Blue Zones, coined by author and researcher Dan Buettner, are places with higher concentrations of thriving older people, many of them centenarians. Those Blue Zones are Ikaria, Greece; Okinawa, Japan; Ogliastra, Greece; Nicoya Peninsula, Costa Rica; and Loma Linda, California. He writes in *The Blue Zones Solution* that one of the top reasons older people are thriving in these areas is the high quality of the food they ingest. He states:

> Most of the Blue Zones' residents I've come to know have easy access to locally sourced fruits and vegetables—largely pesticide free and

205 Environmental Working Group. "EWG's 2019 Shopper's Guide to Pesticides in Produce™." EWG's 2019 Shopper's Guide to Pesticides in Produce | Summary.

206 Charles, Dan. "Are Organic Vegetables More Nutritious After All?" NPR. July 11, 2014.

organically raised. If not growing these food
items in their own gardens, they have found
places where they can purchase them, and more
affordably than processed alternatives.[207]

These populations not only live longer, but they also have
lower rates of disease.[208]

The USDA tests conventional produce every year, and then
the EWG analyzes the numbers and creates a "dirty dozen"
list. The dirty dozen have the highest pesticide residue levels.
While organic is preferable, those on a tight budget should
at least avoid the dirty dozen.

For example, in 2019, avocados tested lowest for pesticides,
while strawberries tested highest. After strawberries, the
highest foods with pesticide residue are spinach, kale, nec-
tarines, apples, grapes, peaches, cherries, pears, tomatoes,
celery, and potatoes. The EWG reports that the USDA found
225 different types of pesticides or pesticide byproducts on
the samples of produce they tested.[209]

---

207 Buettner, Dan. In *The Blue Zones Solution: Eating and Living Like
the World's Healthiest People*, 604–5. Washington D.C.: National
Geographic Society, 2015.

208 Ibid.

209 Environmental Working Group. "EWG's 2019 Shopper's Guide to
Pesticides in Produce™." EWG's 2019 Shopper's Guide to Pesticides
in Produce | Summary.

Kale used to rank lower years ago. In 2019, after testing again, kale rose to number three. The USDA found that more than 92 percent of the kale tested had two or more pesticides. Individual samples of kale had up to eighteen different pesticide residues. The most common pesticide found on the kale was Dacthal (DCPA), which the European Union banned in 2009, and the EPA categorizes as a potential carcinogen but hasn't banned.[210] The EPA states on its website:

> There are no health data on excessive exposure of humans to dacthal or its degradates. In rodents, effects on the liver, kidney, and thyroid were observed following excessive exposure, along with some effects on the lungs.

The website also states that due to runoff, this pesticide ends up in our water supply.[211] The fact that some people are obsessed with kale is potentially problematic. People believe it's a superfood, so they are consuming large amounts of it in salads, smoothies, and juices. Such behavior is counterproductive if they are not consuming organic kale, because they are eating a carcinogen at the same time.

---

210 Ibid.

211 "Summary from the Health Advisory (HA) for Dacthal and Dacthal Degradates (Tetrachloroterephthalic Acid and Monomethyl Tetrachloroterephthalic Acid) ." EPA. EPA, n.d.

Additionally, conventional produce can harm farmers and laborers. At least seventy pesticides used on food in America are potentially or probably carcinogenic.[212] According to the Agricultural Health Study, which was conducted by the National Cancer Institute and the National Institute of Environmental Health Sciences, farmers dealing with these potentially carcinogenic poisons are at risk of developing Parkinson's disease, prostate cancer, asthma, diabetes, and thyroid disease.[213]

Finally, organic food is not sprayed with glyphosate. Chapters 3, 4, and 10 discuss how thousands of people have lawsuits against Bayer, claiming that Roundup gave them cancer. Glyphosate is in Roundup and is sprayed on conventional wheat, oats, lentils, peas, soybeans, corn, flax, rye, buckwheat, millet, canola, sugar beets, potatoes, and sunflowers.[214]

NOT KNOWING PESTICIDES ARE A
THREAT TO OUR FOOD SUPPLY

Last but not least, pesticides are bad news for wild bees. Wild bee populations have been steadily declining for years. From 2008 to 2013, the bee population has declined around 23 percent,

---

212 Batts, Vicki. "Farmers at Higher Risk of Developing Various Cancers Caused by Pesticide Exposure." NaturalNews. August 24, 2016.
213 "News & Findings." Agricultural Health Study.
214 Roseboro, Ken. "Why Is Glyphosate Sprayed on Crops Right Before Harvest?" EcoWatch. January 31, 2019.

and many bee species are facing extinction.[215] Why are the bees dying? While scientists and officials are still debating the cause, the leading hypothesis blames neonicotinoids—a pesticide that is related to nicotine. Neonicotinoids, also known as neonics, can be sprayed on corn, canola, cotton, sorghum, sugar beets, soybean, apples, cherries, peaches, oranges, berries, leafy greens, tomatoes, potatoes, rice, nuts, and wine grapes.[216]

Now you might be thinking, "Yeah, that's sad, but who cares about the bees? They just hover over my food when I'm trying to enjoy a leisurely picnic with my Tinder date." But, bees are a huge part of our global food system. In fact, one third of our food relies on bees or other pollinators to grow.[217]

Pollinators help plants reproduce. Since plants cannot move, pollinators like the bee transfer pollen from one plant to the next. What's pollen? Pollen is plant sperm. You know how bees buzz? They make that sound because they are vibrating on the plant to encourage it to let go of its pollen. In a sense, the bees are making the plant "orgasm"—the buzzing is necessary to get the pollen they need to do their work.[218]

215 Sidder, Aaron. "New Map Highlights Bee Population Declines Across the U.S." Smithsonian.com. February 23, 2017.

216 Grossman, Elizabeth. "Declining Bee Populations Pose a Threat to Global Agriculture." Yale E360. April 30, 2013.

217 Fraser, Carly. "France Becomes The First Country to Ban All Five Pesticides Linked to Bee Deaths." Live Love Fruit. May 13, 2019.

218 Pearson, Gwen. "Bees Are Great at Pollinating Flowers-But So Are Vibrators." Wired. June 06, 2017.

Now you're probably thinking, "Okay, not a big deal. We'll just get some humans or robots to collect the pollen and distribute it. Problem solved." Well, some farmers *do* use vibrators that replicate what the bees can do. Farmers growing tomato plants in greenhouses, for example, do pollinate the crops themselves because they are not outside, so no bees are available to do the work. However, this process is laborious. One farmer attempting to pollinate 640 tomato plants took over a minute to vibrate each plant, which equals nearly twelve hours of labor.[219] That's a long time. Farmers and home gardeners alike can easily access these vibrating devices at home improvement stores. Who knew there were sex toys at Home Depot?

At this point in the book, you won't be shocked to learn that America hasn't banned bee-killing neonicotinoids, while Europe has. As of 2018, the European Union banned all bee-killing neonicotinoids, which include imidacloprid, clothianidin, and thiamethoxam, with the exception of using them in greenhouses.[220] While the EPA reports on its website that scientists are currently reviewing neonicotinoids, there is no ban at all in America in place yet.[221] The EPA is also reviewing two pesticides not mentioned on the EU's pesticide

---

219 Ibid.
220 "Daily News 27 / 04 / 2018." European Commission - PRESS RELEASES - Press Release - Daily News 27 / 04 / 2018.
221 "Schedule for Review of Neonicotinoid Pesticides." EPA. June 06, 2019.

ban announcement—acetamiprid and dinotefuran. In 2018, France was the first country in the European Union to ban five of the neonicotinoids harming wild bees. In addition to clothianidin, imidacloprid, and thiamethoxam, which the EU had already banned, France is also banning thiacloprid and acetamiprid.[222]

Now you might be feeling overwhelmed. Do you need to start your own garden and exclusively eat your own harvest? Never go out to eat again? Stop eating entirely? No. Some food retailers are doing the right thing, like PCC Community Markets.

PCC

Meet Cate Hardy, the CEO of PCC Community Markets, the nation's largest community-owned, organic-certified food market. Created in 1953, PCC now has eleven stores in Washington State and employs over 1,400 people. Cate previously worked at Starbucks in various leadership roles—overseeing retail operations, store and product development, and global supply chain-related work. In 2015, PCC hired Cate as its new CEO. Over the past four years, Cate has helped PCC grow in revenue by 27 percent and net income by 32 percent. Over 95 percent of PCC's produce is organic. This progressive

222 Fraser, Carly. "France Becomes The First Country to Ban All Five Pesticides Linked to Bee Deaths." Live Love Fruit. May 13, 2019.

company gives 50 percent of net income to co-op members and community programs, was the first grocer in America to use BPA-free receipt paper, and follows a triple bottom line model: financial, environmental, and social.

In an interview, Cate tells me how proud she is to work at PCC and how much she admires the trailblazing spirit of the company. PCC featured organic food before it was mainstream. Her focus now is to grow PCC's impact, so customers have access to organic, sustainable products. PCC is community-driven, so instead of just writing checks, the company shows its support actively, whether through having a group of staff members show up at the Seattle Pride Parade or getting a group together to help restore salmon habitats.

Even though PCC is adamant about sticking to the triple bottom line, running a successful grocery store chain in a competitive market and adhering to its values is no easy task. Cate says of the triple bottom line:

> Pretty much everything we do is balanced along those factors. I was just having a conversation with our CFO, and we were talking about the fact that there's an upcoming decision we need to make that throws our environmental bottom line in conflict with our financial bottom line, meaning there is something we would want to

do environmentally, but it's going to cost quite a lot of money and are we prepared to do it. We have those conversations all the time, and sometimes the financial bottom line prevails. Often, the environmental and social bottom line prevails.

PCC is far different from the average grocery store. Besides being environmentally- and socially-minded, the company meticulously screens its products. The company's Quality Standards Committee is responsible for reviewing all products and ingredients that can be sold at PCC. The Committee's standards are high—higher than many grocers in the nation. PCC wants to sell safe and healthy products, but its standards go beyond that. Cate says:

> Safe is the first step for sure, but I think our shoppers are looking for more than that. They are looking for great assurances of responsibility, frankly. In the way things are procured. Whether it's an ingredient or animal welfare policy—what was good enough for us five years ago is not good enough for us anymore. So, we are just constantly raising the bar on ourselves.

Cate tells me PCC also has industry-leading standards for health and beauty products:

Five hundred ingredients that we used to accept into our stores are now no longer accepted, and that was a huge change. We had to discontinue many, many, many products that frankly, people liked, but our research and our perspective on environmental and social, mostly environmental, led us to the conclusion that we could not be consistent with our values and continue to carry these items. We have gone out and found other producers that are producing without these items and brought them into our stores. Some of our former producers are revising their formulas in order to be able to stay on our shelves.

Cate says that the screening process is never-ending—PCC is constantly striving to be better.

PCC is making an impact. PCC not only protects customers directly by curating safe, healthy, and sustainable items, but they also are changing their supply chains, affecting other retailers for the better. For example, PCC sells many natural sodas and used to offer a Coca Cola-like beverage, which contained caramel coloring, a potential carcinogen. PCC refused to continue selling the product. The supplier decided to reformulate the product to exclude the caramel coloring, and many retailers now sell the reformulated product, not just PCC. Because of PCC, less food coloring is being sold.

PCC has also set impressive waste reduction goals. By 2022, PCC plans to eliminate all petroleum-based plastic from the delicatessen. PCC has already ensured that its coffee cups, soup cups, soup lids, cold beverage cups, straws, and cutlery are compostable. Nine PCC stores are LEED-certified, which is a green building certification program. These waste reduction goals are not cheap. Cate says, "Compostables will cost our co-op substantially more than the plastic containers, but will remove eight million pieces of plastic every year from the waste stream." Despite losing money on certain decisions, PCC knows that going plastic-free is the right thing to do.

PCC adamantly believes that organic food is important. Cate says:

> It comes back to our social and environmental bottom lines. There's absolutely no doubt that organic production methods are better for the environment and better for the people who are working in that context. If you're a farm worker and working in a pesticide-laden field, there is no doubt that it's not as good for you as working in an organic field where there's no chemical pesticides. In addition, those pesticides of course have very negative impacts for the environment— whether it's salmon or honeybees—these impacts have been well documented.

Cate is right. Not only are wild bee populations declining because of pesticide use, but the salmon population is also at risk. The pesticides chlorpyrifos, malathion, and diazinon are hurting salmon, which Orca whales eat and absorb, too.[223] *The Seattle Times* reports that the salmon population is dramatically smaller than it used to be:

> Historically, adult salmon returns to the Columbia Basin were at least 10 to 16 million fish annually—today, across the Northwest, less than 5 percent of historic populations of wild salmon and steelhead return to our rivers and streams. Fifteen different salmon and steelhead stocks in Washington state are listed under the federal Endangered Species Act today.[224]

PCC knows that working with organic farmers and selling organic produce is the best way to do social and environmental good.

Has PCC influenced other stores to focus on waste reduction, to protect salmon, to switch to BPA-free receipt paper, and to embrace triple bottom line practices? The company

---

223 Hotakainen, Rob, and E&E NewsJan. "Common Pesticides Threaten Salmon, U.S. Fisheries Agency Concludes." Science. January 12, 2018.

224 Chowder, Duke's Seafood &. "Environmental Impact of Salmon Decline: This Isn't Just about Fish | Provided by Duke's Seafood & Chowder." The Seattle Times. February 07, 2018.

does not know for sure. However, Trader Joe's started using BPA-free receipt paper after PCC did. PCC announced that it would not sell Chinook salmon caught in Washington, Oregon, or British Columbia to protect the endangered Orca whales, which eat salmon as a main food source. A number of restaurants and grocery store chains followed suit. Maybe other retailers were thinking independently about the same topics, but regardless, PCC is glad its practices have the possibility of encouraging other retailers to do the same.

Cate tells me that even if she and her colleagues aren't influencing other stores, they feel content with their decisions. Cate says, "We try to lead the way because we are always looking to be changing and improving our standards. Like I said, what was acceptable to us three, five years ago may not be now, and we are always striving for more."

If you have ever been to a PCC, you'll know that its deli section is impressive, and it prepares dishes from scratch. Cooks even make their own chicken stock. Each store cooks every day using mostly organic ingredients. I ask Cate if she can pick a favorite dish, and she says absolutely not—there are too many dishes she loves. Since the menu changes with the seasons, whenever her favorite dishes return, she is thrilled. PCC is all about transparency—customers can go to the PCC website to find the recipes of their favorite dishes, so they can prepare them at home. The stores even offer cooking classes.

STEPS FOR CONSCIOUS CONSUMERS

1. Buy as much organic food as you can. If you are on a budget, review the EWG website for conventional produce to avoid.

2. To avoid glyphosate, don't buy conventional wheat, oats, lentils, peas, soybeans, corn, flax, rye, buckwheat, millet, canola, sugar beets, potatoes, and sunflowers.

3. Support organic restaurants and farmers' markets in your area.

4. To minimize grocery waste, bring reusable containers and buy products in bulk.

5. Bring reusable bags to the grocery store.

6. Instead of using the disposable plastic produce bags found at most stores, invest in cloth produce bags.

7. Check out warehouse stores in your area. Stores like Costco, which sell more organic food than ever before, reduce the amount of packaging. These options are often cheaper, too.

CHAPTER 14

A PACKAGE TO REMEMBER

———

The way we use the Internet is rapidly changing. Consumers can pretty much order anything in the world with a click of a mouse, shipping is faster than ever, and instant gratification is becoming the norm. As a Seattleite, I see Amazon everywhere—practically everyone I meet works there or knows someone who does. When I drive by the exclusive Amazon spheres, which are part of the Amazon headquarters located in downtown Seattle, I think, "If only Jeff Bezos would let me in. I want to see the conservatory with 40,000 plants," but my second thought should be, "What are the environmental and ethical implications of this type of consumerism? Is it sustainable?"

While only 9 percent of all retail sales in America were online as of 2017, e-commerce sales are rapidly rising and predicted to keep increasing.[225] In 2018, American e-commerce sales rose 15 percent from 2017.[226] Amazon was projected to be responsible for nearly half of all United States-based online retail transactions that same year.[227]

Because Amazon is such a major player in e-commerce, consumers have understandably been critical of Bezos's empire, citing poor working conditions,[228] wasteful plastic packaging,[229] and shipping that's too fast. What's wrong with fast? In 2019, Amazon announced that millions of their products are now eligible for one-day delivery in the United States for Prime members. Faster shipping means an increase in fuel emissions due to inefficient and wasteful delivery routes. The director of global sustainability at the United Parcel Service (UPS), Patrick Browne, says of Amazon's expedited shipping, "I don't think the average consumer understands the environmental impact of having something tomorrow vs. two days

---

225 "United States: E-Commerce Share of Retail Sales 2021." Statista. Statista, n.d.

226 Ali, Fareeha. "US Ecommerce Sales Grow 15.0% in 2018." Digital Commerce 360, February 28, 2019.

227 Laurenthomas. "Watch out, Retailers. This Is Just How Big Amazon Is Becoming." CNBC. CNBC, July 13, 2018.

228 Godlewski, Nina. "Amazon Employees Have Resorted to Urinating in Trash Cans in Some Warehouses." Newsweek. Newsweek, February 26, 2019.

229 Bird, Jon. "What A Waste: Online Retail's Big Packaging Problem." Forbes. Forbes Magazine, July 29, 2018.

from now. The more time you give me, the more efficient I can be."[230] Naturally, other companies have had to compete with Amazon and expedite their shipping, too.[231]

Additionally, one has to wonder whether consumers have become too disconnected and passive when it comes to their spending habits on the Internet. Psychologically-speaking, does the click of a mouse lead buyers to make unwise, impulse purchases? Placing an order on Amazon (and elsewhere on the Internet) feels effortless—a buyer can find what he or she wants in seconds, and after the click of a mouse, the item seems to magically appear on the doorstep. Does this form of consumerism discourage critical thinking more than in-person shopping?

Now, before you set your computer on fire (don't do that—recycle it instead), let's take a breather. I personally love convenience—it not only helps busy people, but it can help disabled people who cannot visit in-person stores themselves. Plus, online shopping can be the greener choice, depending on a complicated myriad of factors, such as the type of packaging used, the speed of shipping, whether the consumer returns an unwanted product, and how far that consumer would have to drive to pick up the item rather than order it online. If only

---

230 DePillis, Lydia. "America's Addiction to Absurdly Fast Shipping Has a Hidden Cost." CNN. Cable News Network, July 15, 2019.
231 Ibid.

there were a way to order items of all categories online and trust that the company's shipping practices, packaging, ethics, and the products themselves were all healthy and sustainable.

EARTHHERO

But wait—such a place exists. Meet Ryan Lewis, founder and CEO of EarthHero, an e-commerce website that sells sustainable and healthy products in all categories, including homeware, clothing, kitchen supplies, technology, pet care, and more. EarthHero has over 100 "sustainability logos"—icons that help categorize the various products it sells. The broader categories include recycled content, upcycled content, organic content, renewable resource, low impact, and responsible.

For example, search for wool dryer balls, one of Ryan's go-to products, which reduces drying time and wrinkles, and seven logos are listed:

- Cruelty free (not tested on animals)

- Sulfate free

- Wool (a natural and renewable resource)

- Sustainable manufacturing (the manufacturer uses partially if not all renewable energy when making the product)

- Safe and fair labor

- Empowers women (brand advocates for gender equality)

- Sustainable lifestyle (the product encourages consumers to live more sustainably)

Ryan tells me that these logos help people easily find what they want according to their values.

EarthHero pays close attention to all aspects of e-commerce other brands might not be considering. At its zero-waste headquarters in Colorado, workers use reusable dishware and compost and recycle as much as possible. The company meticulously screens products to sell and reuses boxes to send orders to customers. EarthHero is the epitome of conscious consumerism.

EarthHero is also a Certified B Corporation, meaning that the company is part of a for-profit network of businesses following triple bottom line practices. EarthHero partnered with 1% For the Planet, an organization established by the co-founder of the famous sustainable clothing company, Patagonia. Members of 1% For the Planet donate at least 1 percent of their revenue to environmentally-focused, non-profit organizations. Lastly, EarthHero is a partner of Carbon Fund, an organization that helps businesses and individuals offset their environmental harm by supporting forestry, renewable energy, and energy efficiency projects. For Ryan, making sure EarthHero had philanthropic DNA from the beginning was important.

Ryan moved to Colorado when he was sixteen and was awe-struck by the beautiful scenery of the state. After college, he backpacked through Europe, Nepal, and Southeast Asia and came to understand the importance of treating our world better. After selling his restaurant supply business, Ryan decided to start EarthHero, determined to create a completely transparent, ethical, and sustainable company.

Ryan was wary of greenwashing, which led him to start the brand in the first place, telling me in an interview, "Brands will say something like, 'This product is made from recycled content.' And you're like, 'Well, how much of it is recycled? What part of it? What do they use?'" Ryan wanted EarthHero customers to feel an instant trust.

He explains why conscious consumerism and buying sustainable and healthy products are so important. He tells me:

> The way we consume everyday products—think about Target or Amazon—you get a new chair for your house or a t-shirt or headphones—the way that thing you bought is made causes a lot of damage along the way to get to you. The way

the materials are extracted, the way materials are manufactured, and then you have to transport it—the way that we consume is damaging.

Ryan understands the intrinsic relationship between health and the environment—everything we apply to our bodies gets in the water and into our world, and everything we release into the world affects us. With conscious consumerism, "there are more trees in the ground, fewer toxins in the air, less trash in the ocean, fewer greenhouse gases being released, fewer developing countries' cultures being destroyed for a quick dollar, and less toxic sludge going into some river."

Ryan continues:

> I think businesses have to create systems and processes to make it easy for consumers to make good decisions. We are the beginning stages of that. I'm seeing it—just with the vendors that we are sourcing—there are more and more companies starting with this in mind. I'm an optimist. I think there will be a new normal, and we will figure out how to exist on this planet without destroying it.

When I ask Ryan, a self-proclaimed minimalist, to identify his favorite EarthHero products, he tells me that some of the products he uses regularly include:

- Bee's Wrap (a zero-waste way to wrap food instead of using tin foil or plastic wrap)

- Dryer balls

- The clothing brands tentree and United By Blue

- SAOLA shoes (made from recycled water bottles, cork, and organic cotton shoe laces)

- Plaine Products (shampoo and conditioner sent in reusable packaging)

Ryan likes incorporating these sustainable and healthy swaps into his life. His daily habits are "trigger points," meaning every time he wraps his sandwich in reusable wrap, for example, he is reminding himself that he cares and wants to make a difference in the world.

Every month, EarthHero adds new brands to its website, and traffic is growing. Ryan knows that his brand is just getting started, and he is excited about the future. He says that he wants EarthHero "to be the go-to household name when it comes to sustainability." Props to Ryan for offering a more sustainable and ethical alternative to Amazon and other e-commerce sites.

STEPS FOR CONSCIOUS CONSUMERS

1. When online shopping, consider choosing a slower shipping option. Do you need that monkey stuffed animal for your niece immediately, or can you wait a few more days? Much like the slow food and slow fashion movements, the slow shipping movement is happening and important to support.

2. Support e-commerce sites that adhere to the triple bottom line.

3. If shopping on Amazon, search for products with "frustration-free packaging." This option ensures that the product will be wrapped in recyclable packaging without excess materials.[232]

4. To reduce your overall waste, Ryan recommends reusable options rather than relying on single-use items. Such products include coffee cups, water bottles, cutlery, and food storage containers. Then, move on to personal care items and see how you can use sustainable and healthy alternatives. He encourages people to look at all items like they look at food labels. Ask, "What is in this item? What isn't in this item?"

---

232 "Frustration-Free Packaging." US About Amazon, May 8, 2018.

CHAPTER 15

IT'S GETTING HOT IN HERE

——

TEFLON

PFAS. No, it's not the name of the next up-and-coming Korean boy band. PFAS, which stands for per- and polyfluoroalkyl substances, is the name of a class of over 6,000 chemicals[233] added to carpets, cleaning products, paint, clothing, food packaging, firefighting foam used in the military, and non-stick cookware.[234] PFAS is even present in certain compostable food containers and cutlery, known as molded fiber

233 "PFAS Master List of PFAS Substances." EPA. Environmental Protection Agency, n.d.
234 "PFAS Contamination of Water." State of Rhode Island: Department of Health. State of Rhode Island, n.d.

products. Molded fiber products include items made of wheat fiber, plant fiber blends, and sugarcane byproducts. When the Center for Environmental Health (CEH), an environmental non-profit organization in California, tested molded fiber products in 2019, they all contained PFAS. PFAS exposure is connected to kidney cancer, testicular cancer, breast cancer, hormone disruption in both adults and children, immune system dysfunction, and thyroid disease.[235]

As farmers use compost to grow food, PFAS is problematic because the food grown in contaminated soil will absorb the chemicals, which we then ingest. Furthermore, PFAS from soil will leach into our water systems.[236]

While many of us try to eat healthy food, cookware is also important, because we are applying heat to corrosive materials. Cookware options seem endless and overwhelming. Do we choose non-stick cookware, copper, cast iron, stainless steel, aluminum, or ceramic? The answer is controversial, but scientific evidence reveals that non-stick cookware is the worst.

Teflon probably sounds familiar to you—most non-stick cookware contains it. The most famous non-stick cookware

---

235 "Healthier Food Serviceware Choices." Center for Environmental Health. CEH, May 2019.
236 Chrobak, Ula. "Eco-Friendly Packaging Could Be Poisoning Our Compost." Popular Science. Popular Science, May 30, 2019.

is made of Teflon, a combination of PFAS chemicals. This cookware is manufactured by Chemours, a company owned by DuPont, a leading American chemical manufacturer.

Teflon has been associated with grave environmental problems. Birds have died from Teflon toxicosis, which occurs when they breathe in non-stick cookware fumes.[237] DuPont has also been named in thousands of claims, which allege that the Teflon manufacturing process polluted public drinking water and air near the DuPont West Virginia chemical plant. Victims have become sick, some even developing ulcerative colitis, pregnancy-induced hypertension, high cholesterol, thyroid disease, and testicular and kidney cancers.[238] PFOA, a type of PFAS previously in Teflon, is so ubiquitous that 98 percent of Americans now have the chemical in their blood.[239]

Perhaps even more disturbingly, DuPont knew about the dangers of PFOA, also known as C8, decades before the public became aware. In a 2013 lawsuit against DuPont, plaintiff Carla Bartlett claimed that PFOA-contaminated water caused her kidney cancer. According to documents revealed in the lawsuit:

---

237 Andrews, David. "Teflon-Killing Canaries and the American Dream." EWG. EWG, May 1, 2015.

238 Kelly, Sharon. "DuPont's Deadly Deceit: The Decades-Long Cover-up behind the 'World's Most Slippery Material.'" Salon. Salon.com, January 5, 2016.

239 "Teflon's Toxic Legacy: DuPont Knew for Decades It Was Contaminating Water Supplies." EcoWatch. EcoWatch, January 4, 2016.

Concerns about the potential toxicity of C8 had been raised internally within DuPont by at least 1954, leading DuPont's own researchers to conclude by at least 1961 that C8 was toxic and, according to DuPont's own Toxicology Section Chief, should be "handled with extreme care."[240]

DuPont internal documents also revealed that in 1961, the company tested rats and rabbits and found that PFOA-exposed animals developed enlarged livers. DuPont also tested human subjects, having them smoke cigarettes containing C8. Smokers would experience flu-like symptoms for hours afterward.[241]

In response to the many lawsuits mentioned above, DuPont reformulated Teflon to exclude PFOA, one of the PFAS chemicals causing significant harm in West Virginia. As of 2019, its website states the company "has moved to next generation technologies . . . [eliminating] the use of PFOA. In 2009 we began to convert customers to the use of these new coatings. Today [the company's] nonstick coatings for cookware and bakeware are made without PFOA."[242]

While PFOA-free cookware is better, Dupont's non-stick cookware may still contain toxic components. Keith Spencer, cover

---

240 Ibid.
241 Ibid.
242 "How Teflon™ Is Made." Chemours. Chemours, n.d.

editor for *Salon* magazine, writes of PFAS chemicals other than PFOA, "there is no evidence these are safe; merely, the fact that they are less-tested means manufacturers can claim ignorance and keep selling their same products with slightly different chemicals."[243] Dr. Tracey Woodruff, director of the Program on Reproductive Health and the Environment at the University of California-San Francisco, adds, "We're in another grand experiment with toxic chemicals, finding out which are bad. They phase one out and then replace it with something not thoroughly tested. It's a sad, never-ending loop."[244]

UNHEALTHY METALS

Additionally, cookware can contain metals that harm human health and disperse into what we cook. Such potentially harmful metals can include copper,[245] iron,[246] nickel,[247] aluminum, arsenic, cadmium, and lead.[248]

---

243 Spencer, Keith A. "The Chemical Industry Doesn't Want You to Be Afraid of Teflon Pans. You Should Be." Salon. Salon.com, February 4, 2018.

244 Ibid.

245 Measom, Cynthia. "What Are the Dangers of Copper Cookware?" Hunker. Hunker, n.d.

246 "Does Cooking with Cast Iron Pots and Pans Add Iron to Our Food?" Go Ask Alice. Go Ask Alice, n.d.

247 Perez, Sandrine. "Is Stainless Steel Cookware Safe?" Nourishing Our Children. Nourishing Our Children, September 28, 2017.

248 OK International. "Cookware made with scrap metal contaminates food: Study across 10 countries warns of lead and other toxic metals." ScienceDaily.

You won't be surprised that heavy metals can cause chaos in the body. Remember President Lincoln, who took mercury antidepressants that made him exhibit rage? Depending on which and how much metal is consumed, one can be fatigued, develop disease or chronic illness, or worse.[249]

Wendy Myers, a California-based functional diagnostic nutritionist and detoxification advocate, says the majority of her clients have mercury, aluminum, and thallium in their bodies, causing chronic fatigue and mitochondrial damage. She states, "Our mitochondria are our body's powerhouses that make our body's energy. Thallium will poison enzymes that transport nutrients into your mitochondria, so that's a big causative factor in people that have chronic fatigue."[250]

In an article posted in PMC, a database run by the United States National Library of Medicine, authors Tchounwou, Yedjou, Patlolla, and Sutton explain how heavy metals enter our environment:

> Although heavy metals are naturally occurring elements that are found throughout the earth's crust, most environmental contamination and human exposure result from anthropogenic activities such

---

249 "Detoxing From Heavy Metals with Wendy Myers." The Energy Blueprint. The Energy Blueprint, August 12, 2019.
250 Ibid.

as mining and smelting operations, industrial production and use, and domestic and agricultural use of metals and metal-containing compounds.[251]

Not all heavy metals are detrimental. For example, in the proper amounts, the body needs iron, zinc, and copper.[252] However, many metals absorbed from cookware are not bioavailable, meaning they have no benefit to the human body on the cellular level.[253]

XTREMA

So, what materials should we be cooking with? Despite transitioning to a healthier lifestyle long ago, I have always been confused about what cookware is best. I have been using stainless steel most of my life, believing it was the healthiest choice. However, after chatting with Rich Bergstrom, founder of Xtrema, a ceramic cookware brand, I am reconsidering that decision. Xtrema's ceramic cookware can be used on top of the stove, in the microwave, and in the oven, all without cracking, which is a difficult feat. Most importantly, Xtrema cookware is safe, passes strict heavy metal extraction testing,

251 Tchounwou, Paul B, Clement G Yedjou, Anita K Patlolla, and Dwayne J Sutton. "Heavy Metal Toxicity and the Environment." PMC. U.S. National Library of Medicine, 2012.

252 Ibid.

253 "Why Ceramic Cookware Is Safer Than Stainless Steel." Xtrema Pure Ceramic Cookware. Xtrema.com, July 26, 2018.

and utilizes a benign manufacturing process that doesn't harm workers or the planet.

Rich tells me in an interview that he has a forty-year background in consumer products. For more than half that time, he worked at Corning Consumer Products in New York, where he learned the ins and outs of ceramics and glass manufacturing. In the late '90s, Corning was sold to a private equity firm, and the company's products were discontinued. Then, in 2004, Rich partnered with a ceramics engineer from Taiwan, who wanted to make ceramic stovetop cookware. Rich used what he had learned at Corning, and by 2008, they rolled out a new product line.

Before launching the product line, Rich and his team, which includes his twin brother and nephew, commissioned extraction testing on his cookware. Such testing is done over a twenty-four-hour period and is an effective way to discover what chemicals or metals might be leaching from a product. He also tested twenty-five metal cookware products from other companies to compare results. A month later, Rich received the results and was shocked. He tells me:

> All of the metal cookware that we tested failed miserably—iron, tin, chromium, nickel, cadmium, cobalt, all kinds of metals. These were not metals that you'd find in food. These are metals you'd find in cookware.

When Rich got the results, he realized so many cookware options on the market were simply not safe, and that the average consumer had no idea. Rich vowed to not only offer a wonderful product, but to make sure his company stood for transparency, awareness, and health. After all, chronic illness, autoimmune issues, cancer, and Alzheimer's disease are on the rise, and the daily lifestyle choices we make add up. Ingesting food contaminated with heavy metals and chemicals three times a day is not optimal. Xtrema tests every batch of products and releases the findings on its website monthly, so consumers can see the data for themselves.

Rich says that Xtrema really gained momentum in 2009, when *National Geographic* magazine released an issue about the greening of America and the importance of being eco-friendly. The issue featured Xtrema. Rich and his team also reached out to health-conscious bloggers who research and recommend safe and healthy products to their readers. These bloggers became affiliates and promoters of the brand.

Xtrema's website also states that the process of making the cookware is pollution-free. Rich tells me that the exterior of a steel factory is dramatically different than one that produces ceramics:

> In the United States, years ago, you couldn't go anywhere in the Northeast without seeing steel

mills. They were everywhere, but with that came a difficult cost. The factories polluted every single body of water on the East Coast. There was a very serious problem with water and soil pollution.

Rich is right. The Environmental Defense Fund, an environmental non-profit advocacy group, produced a pamphlet about the negative repercussions of steel production. The process is energy intensive and polluting. Mining for steel releases dangerous heavy metals and toxicants into the water supply and environment. These toxicants include cyanide, carbon monoxide, benzene, ammonia, and toluene.[254] To address the steel pollution problem, American companies started outsourcing to China, which helped us but didn't help the Chinese people. While Rich's ceramics factory is in China, his factory is significantly less toxic. Instead, he says of the non-toxic ceramic manufacturing process:

> It's clay, it's water, and it's various ceramic minerals that are mixed together like cake batter. In the factory, we mix the dry ingredients together in a container the size of a house. Then, that dry mineral content is mixed with water into a big slurry in another big holding tank, and then that is poured out into the various molds that we have.

---

254 "Buyer Guide: Steel PDF." Environmental Defense. Environmental Defense, n.d.

From there, the product is left to dry for anywhere from twenty-four to seventy-two hours. Then we fire it three times at 2,500 degrees Fahrenheit for twenty-four hours. Almost all of the manufacturing process is handmade by ceramic artisans. We use recycled water, we use natural gas. It takes twenty-two days to make our cookware.

Workers at a steel mill must wear masks. However, in a ceramics factory, unless the worker is mixing the minerals (all soil has naturally-occurring heavy metals in it), there is no reason to wear one. He continues, "By firing ceramics at 2,500 degrees for twenty-four hours, you burn off every impurity that could possibly exist in that clay. That's the whole benefit of ceramics—purity."

Rich transitioned to a largely plant-based diet four years ago. He cooks Asian-inspired dishes with lots of vegetables because he loved the food he has encountered while visiting China for the past twenty years. Naturally, he enjoys cooking these delicious dishes in Xtrema ceramic cookware.

STEPS FOR CONSCIOUS CONSUMERS

1. Inspect the cookware that you use. Research the materials behind every item and then decide what you might want to replace. Avoid non-stick and aluminum cookware at all costs.

2. If opting for stainless steel cookware, be aware that all steel cookware is not made equally. Stainless steel labeled "18/0" is best. The 18 refers to the cookware's chromium content, while the 0 means it is nickel-free. Since nickel has greater health risks, nickel-free cookware is optimal. Examples of nickel-containing stainless steel include cookware labeled 18/8 and 18/10.[255]

3. In additional to traditional non-stick coatings, question newer ceramic non-stick coating technologies. For example, Rich refused to partner with a company selling Sol-gel, a non-stick coating. He told the company, "It took forty years for the public to find out that Teflon was toxic—you have only had your ceramic coatings out for four years. How long is it going to take before you do the proper testing? Why did you bring the coating out so fast into the marketplace?"

4. While minimizing exposure to these metals is best, detoxification and ingesting chelating agents can help the body excrete these metals. A naturopath or doctor in your area can help you do the proper medical testing and get the right treatment. See book Chapters 17 and 18 to learn more about detoxification.

---

255 Perez, Sandrine. "Is Stainless Steel Cookware Safe?" Nourishing Our Children. Nourishing Our Children, September 28, 2017.

5. If you use compostable food containers, they should be PFAS-free. In 2019, Washington State set tighter PFAS restrictions. In addition, brands like Whole Foods and Trader Joe's promised that they will transition to PFAS-free food packaging.[256] When The Center for Environmental Health (CEH), the environmental non-profit organization mentioned previously in this chapter, tested compostable food containers in 2019, researchers there found that the following compostable containers did not contain PFAS: bamboo, clay-coated paper or paperboard, clear PLA (a corn byproduct),[257] paper lined with PLA, and palm leaf.[258]

6. Better yet, bring reusable containers and cutlery with you to the grocery store and restaurants to avoid compostables entirely.

256 Burke, Caroline. "What Are PFAS? Your Takeout Containers Could Have Them & You Wouldn't Know." Bustle. Bustle, August 20, 2019.
257 Grinvalsky, Jim. "Molded Fiber Food Packaging with PFAS: Is It Safe and Compostable?" EBP Supply. EBP Supply, August 16, 2019.
258 "Healthier Food Serviceware Choices." Center for Environmental Health. CEH, May 2019.

CHAPTER 16

BITE ME

—

When I was little, I was always excited to try my friends' toothpastes when we had sleepovers. I was only allowed to use the natural stuff at home, so if I "forgot" my own toothpaste, I could try theirs. At the time, I really wanted white toothpaste with red and green stripes, and the natural alternatives just didn't excite me.

Many consumers are not aware of the toxic and unsustainable ingredients present in the toothpaste they use at least twice a day. According to Vani Hari, food activist and creator of Food Babe blog, the following ingredients should be avoided:[259]

---

259 Hari, Vani. "Is Your Toothpaste Full Of Carcinogens? Check This List..." Food Babe. Food Babe, October 5, 2016.

- Artificial coloring—carcinogenic[260] and may contain heavy metals;[261]

- Carrageenan—carcinogenic and causes gut inflammation;[262]

- DEA (diethanolamine) —when this chemical interacts with other ingredients, it forms carcinogenic nitrosamines.[263] (See Chapter 9 about Sustain Naturals—nitrosamines also can form in the production of condoms);

- Parabens—endocrine disrupters;[264]

---

260 "Summary of Studies on Food Dyes." Center for Science in the Public Interest. Center for Science in the Public Interest , n.d.

261 Ogimoto, Mami, Yoko Uematsu, Kumi Suzuki, Junichiro Kabashima, and Mitsuo Nakazato. "Survey of Toxic Heavy Metals and Arsenic in Existing Food Additives (Natural Colors)." Shokuhin eiseigaku zasshi. Journal of the Food Hygienic Society of Japan. U.S. National Library of Medicine, October 2009.

262 Bhattacharyya, Sumit, Leo Feferman, and Joanne K. Tobacman. "Increased Expression of Colonic Wnt9A through Sp1-Mediated Transcriptional Effects Involving Arylsulfatase B, Chondroitin 4-Sulfate, and Galectin-3." Journal of Biological Chemistry. Journal of Biological Chemistry, June 20, 2014.

263 "NITROSAMINES." EWG's Skin Deep Cosmetics Database. EWG, n.d.

264 Darbre, Philippa D, and Philip W Harvey. "Paraben Esters: Review of Recent Studies of Endocrine Toxicity, Absorption, Esterase and Human Exposure, and Discussion of Potential Human Health Risks." Journal of applied toxicology : JAT. U.S. National Library of Medicine, July 2008.

- Preservatives, such as DMDM hydantoin, diazolidinyl urea, imidazolidinyl urea, polyoxymethylene urea, methenamine, quaternium-15, sodium hydroxymethylglycinate, 2-bromo-2-nitropropane-1,3-diol (bromopol), 5-bromo-5-nitro-1,3 dioxane (bronidox), and glyoxal—these preservatives all deliver formaldehyde, which is carcinogenic;[265]

- Glycerin (GMO soy, cotton, or canola oil), citric acid (GMO sugar), xanthan gum (GMO sugar), xylitol (GMO corn), and lecithin (GMO soy) —potentially carcinogenic;[266]

- PEGS (polyethylene glycols) and Propylene Glycol— commonly contaminated with 1,4-dioxane, which is a carcinogen;[267]

- Sodium Laurel Sulfate (SLS) and Sodium Laureth Sulfate (SLES) —also commonly contaminated with 1,4 dioxane.[268] These ingredients also irritate the lining of the mouth, causing users to absorb more chemicals.[269]

---

265 "FORMALDEHYDE." EWG Skin Deep Cosmetics Database. EWG, n.d.
266 Cassidy, Emily. "Did You Know That Monsanto's Glyphosate Doubles the Risk of Cancer?" EWG. EWG, October 6, 2015.
267 "1,4-DIOXANE." EWG Skin Deep Cosmetics Database. EWG, n.d.
268 Ibid.
269 Herlofson, BB, and P Barkvoll. "Sodium Lauryl Sulfate and Recurrent Aphthous Ulcers. A Preliminary Study." Acta odontologica Scandinavica. U.S. National Library of Medicine, October 1994.

- Triclosan—researchers discovered that mice that ingested triclosan had severe gut inflammation.[270] While the FDA banned triclosan from antibacterial soap in 2016,[271] it is still allowed in toothpaste and other products.[272]

Think we don't absorb toothpaste since we don't deliberately swallow it? Consider this medical study published in the International Journal of Pharmacy and Pharmaceutical Sciences in 2011. The article states, "Drug delivery via the oral mucous membrane is considered to be a promising alternative to the oral route."[273] In other words, the mouth has the ability to effectively absorb substances even if not swallowed.

Researchers have also found that people swallow 5-7 percent of any toothpaste they use. If you use a toothpaste with questionable ingredients, you could be consuming these chemicals

270 Healy, Melissa. "Triclosan Could Be Really Harmful to Your Gut, and It's Probably in Your Toothpaste." Los Angeles Times. Los Angeles Times, May 31, 2018.

271 "FDA Issues Final Rule on Safety and Effectiveness of Antibacterial Soaps." U.S. Food and Drug Administration. FDA, September 2, 2016.

272 Healy, Melissa. "Triclosan Could Be Really Harmful to Your Gut, and It's Probably in Your Toothpaste." Los Angeles Times. Los Angeles Times, May 31, 2018.

273 Narang, Neha, and Jyoti Sharma. "SUBLINGUAL MUCOSA AS A ROUTE FOR SYSTEMIC DRUG DELIVERY." Innovare Academic Sciences, 2011.

every day via the membranes in your mouth *and* your digestive tract.[274]

Furthermore, a report published by Cornucopia Institute, a non-profit sustainable agriculture advocacy group, states that the FDA does not regulate toothpaste since it is considered a cosmetic,[275] and as you'll recall, only a small number of cosmetic ingredients are banned in America.

Fluoride is also a controversial ingredient added to the majority of toothpastes. In addition, fluoride is added to much of the American water supply to protect dental health and prevent tooth decay. While the FDA regulates fluoride in dental products and mandates that fluoride be disclosed,[276] it may not be as beneficial as many believe. Look at the back of any conventional toothpaste, and you'll see that calling poison control is recommended if children under six eat it. Enamel mottle, also known as dental fluorosis, afflicts children who ingest too much fluoride—their teeth become discolored and damaged.[277] The Cornucopia report writes of fluoride:

---

274 Ciancio, Sebastian G. "Baking Soda Dentifrices and Oral Health." The Journal of the American Dental Association. JADA, November 2017.

275 Cornucopia. The Cornucopia Institute, 2016.

276 "CFR - Code of Federal Regulations Title 21." FDA. FDA, n.d.

277 Cornucopia. The Cornucopia Institute, 2016.

One of the main concerns with fluoride is its potential chronic toxicity. Fluoride accumulates in bones, which can lead to a condition called skeletal fluorosis, characterized by reduced flexibility, chronic joint pain, arthritic symptoms, and bone fracture.[278]

In 2012, Harvard University released a meta-analysis which surveyed many studies of the neurodevelopment of children who drank fluoridated water. The researchers found "strong indications that fluoride may adversely affect cognitive development in children. Based on the findings, the authors say that this risk should not be ignored, and that more research on fluoride's impact on the developing brain is warranted."[279]

Lastly, conventional toothpaste has the added disadvantage of contributing to the plastic pollution crisis described in Chapter 7. Many of us do not think about plastic when it comes to our oral care routines. Did you know that one billion plastic toothpaste tubes are thrown into landfills each year? That's enough to fill the Empire State Building fifty times.[280]

So, what can be done?

---

278 Ibid.

279 "Impact of Fluoride on Neurological Development in Children." Harvard T.H. Chan. Harvard , July 25, 2012.

280 Limitone, Julia. "Vegan Toothpaste Pill Aims to Cut Plastics in Landfills." Fox Business. Fox Business, February 18, 2019.

BITE

Meet Lindsay McCormick, founder of Bite Toothpaste Bits, a zero-waste and all natural way to brush your teeth. Bits are tablets that look just like mints. You pop one into your mouth, bite down, wet your toothbrush, and brush your teeth normally. The tablets dissolve and foam effortlessly. Instead of a plastic tube, the bits come in a cute little glass jar with an aluminum lid. For future orders, you subscribe and Bite sends you refillable bits in compostable packaging. You pour the new tablets into the original glass jar. You can even take them through airport security because they are liquid-free.

Lindsay never expected she'd become an entrepreneur, but she was environmentally-focused since adolescence. As a teen, when her parents asked her what she wanted for her birthday or Christmas, she'd ask to donate money to organizations that allowed you to "adopt" an endangered animal in the wild. In college, she wrote her senior thesis on the importance of biomass and the deforestation of the rainforest.

Fast forward to adulthood, and Lindsay was a TV producer for a reality show called *House Hunters*. She was flying frequently, using travel-sized plastic toothpaste tubes. Lindsay wanted a more sustainable toothpaste option. However, she couldn't find a toothpaste that was both sustainable and contained non-toxic ingredients. Lindsay decided to make her own.

Lindsay took online chemistry classes, invested in a tableting machine so she could develop her product at home, and opened an online shop. Zero-waste bloggers and vegans found her product, which propelled Bite forward. Then, in 2018, *Women's Health* magazine asked her to collaborate. Lindsay shot a video of herself using the product before work, the video went viral, and Bite was officially on the map. *Cosmopolitan* magazine, *Seventeen* magazine, *Oprah* magazine, *Buzzfeed* online magazine, and Goop (Gwyneth Paltrow's brand) endorsed the product.

In an interview, I ask Lindsay how people react to her product. After all, the idea of chewing on a tablet instead of using regular toothpaste is unexpected for many people. She says that, at first, when she'd talk about the product with friends, they didn't immediately understand. Now things are different:

> When I tell them about how they are toothpaste tablets and how our mission is to get one billion plastic toothpaste tubes that end up in landfills and oceans out of the equation, they get it. It's because we are finally understanding that we need to make some big changes in our routines. Obviously, it's a whole new form factor, and there is nothing more ingrained in us than brushing our teeth.

When designing Bite, Lindsay wanted to make sure that her products would be easy for people to integrate into their lives. She says:

> For some people, they will say, "Okay, doing something in a more sustainable way is incredibly important to me, and I'm okay with being inconvenienced for this," and some people will do that and that's amazing. But a lot of people, they can't or they won't. It was really important to make this something that most people would like as much or more than conventional toothpaste, so making a sustainable choice didn't feel like a burden.

Lindsay wanted switching away from conventional toothpaste to feel comfortable, fun, and not serious.

Lindsay is also very passionate about safe ingredients, because she thinks it's vital to know what we are putting in our bodies and down the drain. Every one of her ingredients has a good score on the Environmental Working Group database. She adds, "We're really big on iteration, so we just did an overhaul of all of our ingredients to go 100 percent palm-oil free." Palm oil contributes to deforestation, jeopardizing endangered animals' habitats, in addition to causing pollution.[281]

---

281 "Palm Oil." WWF. World Wildlife Fund, n.d.

Lindsay decided to replace the ECOCERT palm oil Bite had been using with Sodium Cocoyl Isethionate, which is a safe and mild ingredient derived from coconut oil.

I ask Lindsay about high price points of eco-friendly and healthy products. She tells me that before starting Bite, she didn't understand why sustainable and healthier products were often more expensive. She says:

> For us, every step of the way, we are faced with the fast and cheap decision or the sustainable one that unfortunately also ends up being more expensive. From using xylitol and erythritol as opposed to sorbitol, which would be so much cheaper, or using glass bottles, which is so much more expensive than plastic. Plastic would cost pennies. Some of the issues will be solved as our business grows. When you hit certain tiers of scale, things get cheaper. Your ingredients may get cheaper. But at the same point, other things might get more expensive.

She shares that, for example, as her business grows, she might need to hire an engineer to run her website because there are so many more views. Additional expenses appear with growth. She continues:

When it comes to right now, what we are up against is that the entire supply chain and business in general is set up to be as cheap and crappy as possible, cost effective as possible, even at the expense of our health and the planet. Because of that, when you try to do something different, you pay a premium every time. It's crazy.

For example, Lindsay did not want plastic laminate on Bite's label, so she had to pay more to exclude the laminate. Nevertheless, she plans to lower costs as much as she can, so Bite can be accessible to everyone.

Lindsay also feels that while consumers may have to spend more money on certain sustainable and healthy products, they can also reduce their spending on others. For example, Lindsay makes her own facial toner out of apple cider vinegar and water, which is considerably cheaper than what is available at the store. Another great example of reducing spending ultimately is investing in reusable menstrual products, as explained in Chapter 10.

Bite offers two different kinds of toothpaste—Fresh Mint and Fresh Mint with Activated Charcoal (charcoal is a natural teeth whitener). Bite also is part of the slow shipping movement, which deliberately slows delivery to minimize the company's carbon footprint.

Lindsay saw a problem in her own life and invented a solution. Now, thousands of people wake up every morning, and instead of squeezing toothpaste out of plastic tubes that will end up in landfills, they each simply pop a bit into their mouths and get on with the day.

STEPS FOR CONSCIOUS CONSUMERS

1. When toothpaste shopping, research every ingredient and think about the end life of the packaging. Will it end up in a landfill? Can it be recycled? Is it compostable?

2. Consider using a compostable bamboo toothbrush.

3. Make sure your dental floss is PFAS-free. Products like Oral B Glide contain PFAS. In 2019, researchers found that women who used Oral B Glide dental floss had higher levels of PFAS in their blood than those who did not.[282] Instead, go with floss coated in natural wax. Tom's of Maine, Desert Essence, and Eco-Dent make safer flosses.

---

282 Boronow, Katherine E., Julia Green Brody, Laurel A. Schaider, Graham F. Peaslee, Laurie Havas, and Barbara A. Cohn. "Serum Concentrations of PFASs and Exposure-Related Behaviors in African American and Non-Hispanic White Women." Nature News. Nature Publishing Group, January 8, 2019.

CHAPTER 17

ARE YOU DETOX CURIOUS?

———

Detox. Many associate the word with drug users who start intensive programs at clinics and hospitals to ease withdrawal symptoms as they quit addictive drugs. Others think of detoxification teas (chemical- and sugar-laden laxatives) promoted by Instagram influencers. Still others immediately associate the term with "quackery," "pseudoscience," "scam," and "marketing ploy." Just Google "detox," and many articles pop up, claiming that the organs of the body are already detoxing and no further assistance is needed.

However, detoxification is nothing new and has been a part of various cultures for hundreds, even thousands, of years. It's not a scam or a marketing ploy. Yes, some companies take advantage of the word detox and sell unhealthy products that we should not be consuming. But the concept of detoxification, when done correctly, is one of the healthiest human activities.

Other cultures use detoxification as an integral part of their medicinal arsenal. For example, some Native Americans use a sweat lodge, a version of the sauna. People go into the lodges to physically and spiritually purify themselves.[283] Additionally, Ayurveda, the ancient Hindi medical tradition, utilizes detoxification. One of the most practiced Ayurvedic rituals is called panchakarma, which is a five-step detoxification program in which certain types of massages, enemas, and steam baths are administered, in addition to the ingestion of healing and detoxifying foods.[284]

The articles and people discounting detoxification disregard the unfortunate fact that we live in an increasingly toxic world. In fact, ten million tons of toxic chemicals are released into our environment each year—that is over 21 billion pounds.[285] Add up the average person's exposure to

283 Gadacz, René R. "Sweat Lodge." *The Canadian Encyclopedia*, The Canadian Encyclopedia, 7 Feb. 2006,
284 Rose, Sahara. "What's a Panchakarma? Panchakarma Ayurvedic Detoxification Treatments Explained." Eat Feel Fresh. December 15, 2016.
285 "Toxic Chemicals Released by Industries This Year, Tons." Worldometers.

chemicals in conventional personal care products, the pesticides and herbicides in non-organic food, the heavy metals, chemicals, and pharmaceutical residue in our water supply, and the pollutants in the air, and one can easily conclude that our bodies were not prepared for this kind of onslaught. Our bodies accumulate toxins—whether they be in the umbilical cord of an unborn baby or in fat.[286]

The aftermath of the attacks on the World Trade Center on September 11, 2001, offer insight into the power and efficacy of detoxification. After the destruction of the Twin Towers, emergency responders and other workers breathed in carcinogenic, immune-impairing, and disease-causing agents, such as asbestos, radionuclides, benzene, dioxins, polychlorinated biphenyls (PCBs), fiberglass, mercury, lead, silicon, and sulfuric acid. Many workers became sick, facing respiratory, mood, and gastrointestinal issues, among other symptoms.[287]

To address these workers' acute and debilitating symptoms, the New York Rescue Workers Project was established. The program helped people excrete toxicants from their bodies

---

286 Lee, Y.-M., K.-S. Kim, D. R. Jacobs, and D.-H. Lee. "Persistent Organic Pollutants in Adipose Tissue Should Be Considered in Obesity Research." Obesity Reviews. December 02, 2016.

287 Cecchini, Marie A., David E. Root, Jeremie R. Rachunow, and Phyllis M. Gelb. "Health Status of Rescue Workers Improved by Sauna Detoxification." Arthritis Trust. Arthritis Trust of America, April 2006.

via an intensive detoxification program, which included daily exercise, sauna therapy, and nutritional supplementation.[288]

The organization tracked 484 patients who finished the program. All had fewer sick days and reported a dramatic reduction in the severity of reported symptoms. The workers reported improvement of problems with skin, heart, lungs, ears, nose, throat, brain, gastrointestinal, musculoskeletal, and immune system. They also reported improvement in emotional and cognitive issues. Their balance improved. The majority were able to stop taking medications to treat their symptoms. For those who were struggling with thyroid issues, the majority returned to normal functioning. Detoxification helped these workers gain back their health. If they had not taken part in the program, toxicants would have remained in their bodies for years.[289]

Thankfully, though the average person is not exposed to the toxicant levels of post-9/11 workers, detoxification can be powerful and life-changing. Since our world is increasingly full of toxic chemicals, detoxification is a smart practice and even a prerequisite for good health and longevity.

---

288 Ibid.
289 Ibid.

TUMMY TEMPLE

Meet Kristi and Tim Zimmer, wife and husband co-founders of the Tummy Temple, two detoxification centers in Seattle and Olympia, Washington. The Tummy Temple offers various treatments such as Chi Nei Tsang, a type of Taoist massage, and lymphatic drainage, a massage that helps move stagnant lymph in the body. The clinic provides craniosacral therapy, a type of cranial manipulation; colon hydrotherapy, a cleansing practice to remove toxins; homeopathy; nutritional and naturopathic counseling; and Reiki, a form of energy medicine. The Tummy Temple also has its own line of supplements and recently launched a non-psychoactive line of Cannabidiol (CBD) products called Bliss Biologics.

Kristi's upbringing was very different from Tim's. Kristi grew up on a farm in rural North Carolina until she was around eight years old. While farm life made her feel connected to the earth, her family used pesticides and she was a junk food fanatic throughout college, knowing nothing about natural health. Her family had a history of cancer and illness, and Kristi was experiencing gastrointestinal issues, headaches, chronic fatigue, and inflammation on a daily basis for years. A vegan friend kept insisting she read a book called *Diet for a New America,* by John Robbins. After finally reading it, the book opened her eyes. She tells me in an interview:

I became a vegetarian as I was reading it, and it just kind of woke me up to a lot of things I had not considered before. It made my whole world make sense, because, again, I was kind of raised in this natural order on this farm, but then everybody got sick. So, then I started to learn these are all the reasons why. And I got actually pretty angry initially, because I lost a lot of family members to cancer and illness, and it seemed so avoidable after digging into information.

She couldn't go back to how she'd been living before. Kristi continued to read and educate herself. She decided to move to Seattle, a hub of alternative health. She opened a small practice at the Evergreen Integrative Medicine Center with two other practitioners. There, she provided nutrition counseling, massage, and colon hydrotherapy.

Tim, on the other hand, had health-conscious parents who did not allow junk food in the house. He tells me that his father started researching gluten, which was considered a healthy alternative superfood at the time. He would make Tim gluten burgers, which is practically sacrilegious in the alternative health world today.

Kristi and Tim met on a blind date and got engaged twelve days later. Tim tells me, "When I met Kristi, I was exposed to a whole new realm of natural healthcare, and that was something that had not previously been part of my life." They decided to work together, and discussed their vision for a detoxification clinic during many long walks.

Kristi and Tim opened their clinic next door to where Kristi had been working. They wanted to create a healing environment that was less focused on illness and more oriented toward sacredness, recovery, and lightness. Kristi says of the name, Tummy Temple:

> I felt like using a word like tummy, which has been kind of controversial in our business, sort of like, "Will you take that seriously?" I felt like it needed to be that light and innocent in order for people to let go of some heavy stuff. So the name kind of arose from this feeling of let's lighten it up a bit, let's make it a sacred experience, a sacred space.

As soon as Kristi and Tim launched the Tummy Temple, they knew they were making an impact. Tim reminisces:

> I remember when we opened the Tummy Temple, it was in the first week and I was outside the front door, and I was potting a plant as one of the clients left, and it was one of the first clients we

had at the Tummy Temple. This woman opened the door and she looked at me, and her eyes were all watery, and she just said, "Thank you so much. For the first time in years, I feel my life force coming back to me."

Tim remembers her as glowing. That interaction was all he needed to affirm that he was pursuing the right profession— helping people and giving back. Tim finally found professional fulfillment co-founding The Tummy Temple, which he says felt so different from previous jobs he had held.

Kristi adds regarding the glowing woman, "You know that woman who came out crying, it could have been the first bowel movement she had had in a week [from colon hydrotherapy]—it could have been that simple." Tim agrees, and they laugh. She continues, "Or, it could have been a severe trauma—just being able to cry and let that out." Tim chimes in:

A lot of the work that we do helps people regain hope. A lot of people get to a point of feeling hopeless and frustrated, and they don't know what to do, and they don't know how to get out of the situation they are in, and they don't know what better looks like. People are just told that it's all in their heads.

Kristi and Tim describe the differences between conventional and alternative medicine. Kristi says:

> I always have to say that conventional medicine is designed to keep you alive. It's really emergency care, and once disease sets in, how long can we keep you alive. It's not preventative, at least not yet. In general, even a lot of the things that we are talking about that are affecting quality of life, they don't fall into the realm of you have a disease and you're about to die.

She says that if a person has a chronic migraine, it affects every aspect of that person's life but isn't a life or death situation. Often doctors will just prescribe pills to treat the symptom. She admires natural medicine because it searches for the root cause.

Kristi discusses how the Tummy Temple has helped transform the health of clients. She says:

> I always look at quality of life and stories like somebody having insomnia for ten years and basically becoming disabled from that and just couldn't figure it out and with the right nervous system support, calling up two weeks later, and saying, "I'm sleeping for the first time in ten years,

and I've got my life back." And then a month later, back to work.

She tells me that they just had a client who had a headache every single day, and via detoxification, craniosacral therapy, and supplementation, her headaches had reduced to once a month. She says, "You cannot underestimate that! That's like a life reborn." Tim adds:

> My favorite stories are always the ones that are the life changing—the people who are suffering from health issues that prevent them from dating, you know, from just going out in public, from working, from walking, from being functional, from having babies, and miscarriage after miscarriage and finally being able to carry a baby. There are so many of those life-changing shifts that people have. Those are the ones that are so gratifying.

Kristi and Tim are aware that we are exposed to more toxicity than our bodies were designed to handle. They recognize that The Tummy Temple is needed in a polluted world. Kristi hopes that, as people heal through the services offered there, they will live more natural lives. Kristi says:

> When I first started in this business over twenty years ago, I was in this little Seattle bubble with

a few doctors who were really digging into environmental medicine and starting to wake people up, and they had these mystery diseases and they didn't know where their complications were coming from. These pioneering doctors who were starting to examine this toxicity issue and starting to prove, "Oh yeah, this person has this much heavy metals in their body or this much petroleum." And they would pull these substances out of people, and their health would return. And that was radical.

Tim adds that our stress levels are also higher than ever before because of the fast-paced world that we live in, and we are numbing that stress with processed foods, alcohol, and drugs. He says, "Toxins are instigating inflammation and pain and foggy brain and all kinds of symptoms that are uncomfortable and make it harder to make an enjoyable life and harder to live longer."

Fast forward to 2019: The Tummy Temple has served over 20,000 clients in seventeen years of business and now employs over twenty people. Kristi and Tim tell me that people come to the Tummy Temple because of the unusual services it offers and recommendations from clients who have benefitted from their services. Kristi says people are attracted to the Tummy Temple because:

We present vitality. We have that tuning fork of people who want to let the illness go and be well. They want to work on their illness, but they want to work on it in terms of really focusing on the being well part. Again, the whole environmental piece of going into a clinical setting and going to work on these things but also having an environment that really brings out your vitality. You kind of forget for that hour or two hours that you're there that you're sick. People need that.

Tim agrees:

> I think a lot of people are not necessarily feeling that good and don't know how to feel better, and they are looking for that guidance. And a lot of people want to feel better, but they don't know what that looks like. They are confused and are looking for that vision. The Tummy Temple practitioners provide that, as much as in services, but also in the lifestyle that we all lead as a community. There is a lifestyle that is appealing to folks, it's intriguing and inspiring.

Unlike many clinics, the Tummy Temple does not feel impersonal. Instead, clients walk under a torii gate before entering the building to indicate that they are about to enter a sacred space. A torii gate is a traditional Japanese gate that marks Shinto shrines. Inside, the atmosphere is quiet and relaxing. Asian-inspired art adorns the walls, cell phone use is discouraged, and delicious samples of healthy snacks are often available at the front desk. The Seattle location also offers regular yoga classes, sound bowl ceremonies, and wellness education classes.

Kristi and Tim stand for a type of medicine in which symptoms have root causes and can be reversed (not just suppressed) with the right treatment. This commonplace and seemingly obvious idea in the alternative health world needs to be incorporated into mainstream medicine. Thanks to the Tummy Temple, people are learning about the essential practice of detoxification and implementing it in their own lives.

STEPS FOR CONSCIOUS CONSUMERS

1. If you're detox curious, try it out and see if it makes you feel better. Find a detoxification center in your area.

2. Examples of detox practices you can try from home include: drinking lemon water, using a sauna, jumping on a trampoline for lymph drainage, applying castor oil packs to your abdomen, taking Epsom salt baths, and dry skin brushing.

CHAPTER 18

YOU MAKE ME SWEAT

———

Chapter 17 discusses the health problems caused by toxic chemicals and how detoxification can address them. In addition to the services offered by clinics like the Tummy Temple, saunas can provide an efficient and relaxing way of detoxing from home.

In 2015, Dr. Jari A. Laukkanen, professor at University of Eastern Finland in Kuopio, investigated the health of over two thousand Finnish men between the ages of forty-two to sixty over a twenty-one-year period. He came to the overall conclusion that the higher the frequency of sauna use, the lower the risk of fatal cardiovascular disease, sudden cardiac death, and mortality from all other causes. The risk of sudden cardiac death was even lower for men who spent a greater time in the sauna.[290]

As of 2013, Finland had 5.3 million people and 3.3 million saunas.[291] The Finnish people love their saunas so much that a Helsinki Ferris wheel even has a sauna in one of the cabins.[292]

Many sauna brands are available. Which brand should you trust? Which sauna is best? Do you have to go to Helsinki?! No, not to worry.

SAUNASPACE

Meet Brian Richards, founder of SaunaSpace, which produces near-infrared saunas that utilize incandescent light. In this type of sauna, users sweat from the heat, which is an extremely effective way to detoxify the body. They also receive benefits from the light itself. The incandescent lights are a form of light therapy, also known as photobiomodulation. Photobiomodulation is "the utilization of non-ionizing electromagnetic energy to trigger photochemical changes

---

290 Laukkanen, Tanjaniina, Hassan Khan, and Francesco Zaccardi. "Association Between Sauna Bathing and Fatal Cardiovascular and All-Cause Mortality Events." JAMA Internal Medicine. American Medical Association, April 1, 2015.

291 Bosworth, Mark. "Why Finland Loves Saunas." BBC News. October 01, 2013.

292 Julia, Camille. "The Ultimate Helsinki Sauna Crawl." Becoming Fully Human. March 01, 2019.

within cellular structures that are receptive to photons."[293] In other words, it's light that can heal.

Brian didn't know he'd end up in alternative health—in fact, he majored in chemistry and Spanish and thought he'd follow in his parents' footsteps by becoming a doctor. However, when Brian sat across from a Vanderbilt University medical school admissions officer, he felt so uncomfortable that he ended up leaving in the middle of the interview. Brian did not want to be part of a system that promoted pharmaceutical treatments rather than root-cause solutions. He tells me in an interview, "I didn't want to be part of this framework. It's not where I can bring the most value."

Brian started sauna therapy because he was experiencing health issues, including adrenal fatigue, unusual acne on his torso, and insomnia. His doctors recommended Accutane and other pharmaceutical treatments, but he knew he wanted to pursue natural approaches if possible. He tells me:

> I wanted an approach that addressed the root cause, so I kept coming back to toxicity and the use of saunas in many human cultures and traditions that go back thousands of years to clean out the body. I actually found what is the

---

293 "What Is Photobiomodulation? Photobiomodulation Process." Vielight Inc.

incandescent sauna, so it's not just an infrared sauna. It uses the incandescent light bulb, which is a unique light source, so it is infrared but it's not like a typical infrared sauna out there. It's primarily emitting what's called near-infrared wavelengths, and those are the magical infrared wavelengths—those are the ones that stimulate mitochondrial healing systems.

Brian discovered two resources that would change his life forever. The first was a book published in 2003 by Dr. Lawrence Wilson, *Sauna Therapy for Detoxification and Healing*. In the book, Wilson introduced the benefits of detoxification through sauna therapy. Wilson also wrote about a version of a sauna that Dr. John Harvey Kellogg (*yes*, Kellogg's, as in the brand that makes breakfast cereal) invented in 1891: the electric incandescent bath. The electric incandescent bath was a form of near-infrared photobiomodulation.

Brian knew that he needed to try a near-infrared sauna. Unfortunately, no such product existed in 2008, so he decided to construct his own. He used chicken wire, plumbing pipes, and hardware cloth to make his first sauna. Today, SaunaSpace saunas are based on that prototype. Instead of the average sauna that looks like a wooden phone booth with a glass door at the front, SaunaSpace saunas look like tents, weigh under 100 pounds (depending on the model), and are portable. Customers can set them up in minutes and store them when not in use. SaunaSpace also sells light panels, which means customers can convert their existing wooden saunas with SaunaSpace incandescent technology, or they can buy the light panels for targeted use, such as pointing them at a sprained wrist or blemish.

After several sessions with his homemade sauna, Brian's insomnia went away. Months later, use of the sauna forty minutes a day, four to five days a week, resolved his adrenal fatigue, acne on his torso, and other issues. He tells me, "I was less irascible, I was more able to concentrate, more patient, I had much more energy to do the things I wanted to do." Brian did not realize how good he could feel, because he had felt subpar for so long. He attributes his former symptoms to toxicity:

> Now I see today that it's not the only factor in disease, but I would say it's a core factor, one of the primary factors in the function that leads to a diseased state for humans. Exposure to modern toxicants—petrochemicals, plastics, and heavy metals, which are a big part of modern life and modern technology, and there are other things, too. We don't have an experience with natural light like we used to. We have other unnatural things of modern life that have gotten us away from our ancestral experience.

Brian enthusiastically shares that while the sauna has been an essential part of various cultures for thousands of years, detoxification is:

> more crucial than ever because our exposure to toxins is greater than ever. It's really not just another therapy. Nowadays, it's an essential aspect of keeping oneself healthy because you can't avoid the exposure, even if you try. Like, if you look at glyphosate, Roundup, even in certified organic products, it's likely to have some kind of glyphosate contamination because it's a water-soluble chemical, and it's in everything.

As Brian continued to detox the chemicals, heavy metals, and toxins in his body and began to feel better, he saw how much the sauna had helped him and wanted to share that product with others. He used the same prototype and built saunas for his friends, family, and acquaintances, but his main focus was his regular gig—remodeling houses.

Then, Brian formed a relationship with Dr. Wilson, whose book, *Sauna Therapy for Detoxification and Healing*, had helped him so much. Dr. Wilson asked Brian if he would like to sell his saunas on Dr. Wilson's website, and Brian eagerly accepted.

Brian started selling four or five saunas a month and began developing new designs and concepts. He wanted to expand his business, so at the end of 2013, he convinced a bank to give him a small business loan, bought the necessary machinery to build his saunas, hired his first employee, and secured a workshop. Before long, the business had grown to twenty-five employees and a 17,000 square-foot space that he tells me the company will soon outgrown.

Brian says that you have to have the right personality type if you're starting a company: "You have to be the entrepreneur, dreamer, manager, and technician at the same time. In one day, in eight hours, I'll jump and do twenty or thirty different things from design to marketing to production to legal and finance, you name it."

By healing himself, Brian was able to help thousands of people all over the world through SaunaSpace—even the founder of Twitter has a SaunaSpace sauna. SaunaSpace is encouraging people to take time for themselves and relax, rejuvenate, and detoxify.

STEPS FOR CONSCIOUS CONSUMERS

1. If you can't get your own sauna, go to a local gym, spa, or detox center to sweat.

2. Taking binders, such as activated charcoal or chlorella, fifteen minutes before sweating can help mop up toxicants excreted during a sauna session.

3. To replenish electrolytes after sweating, drink coconut water or water with added minerals or salt.

4. Don't just go online and buy the cheapest sauna you can find. When materials are heated up, they offgas, so the materials of a sauna must be non-toxic so that you do not absorb additional toxins when detoxing. SaunaSpace saunas are made with hypoallergenic, non-offgassing materials primarily sourced in America.

5. Electromagnetic field (EMF) levels found in saunas are also important. You want a sauna with low EMF levels, because you want to heal in a sauna, not make yourself sicker. Brian says, "We have completely mitigated electromagnetic stress from the sauna experience, which is pretty unique in the sauna world. We are the first company ever to come out with a faraday cage sauna." A faraday cage is an enclosed space that has the ability to block EMFs.

PART 3

LEADERS IN CONSCIOUS CONSUMERISM

CHAPTER 19

DON'T BE TRASHY

———

What about individuals who are making a difference without selling a product or service? Heather Trim, Rob Greenfield, and Shelbi Orme are three inspiring people who have dedicated their lives to sustainability, health, and conscious consumerism.

Name: Heather Trim

Current location: Seattle, Washington

Known for: Executive director of non-profit organization, Zero Waste Washington

Her background: Heather has worked in environmentalism for decades. Growing up in Texas, Heather remembers when she fell in love with geology in ninth grade and knew then and there that she wanted to work in the sciences. She tells me in an interview, "I had this teacher who came in and taught us geology. She came from the oil and gas industry, and she made it really fun—kind of like doing geology is like doing detective work."

In college, Heather majored in geology and then went to graduate school for geochemistry. Post-grad, she worked for the state-level EPA in California, known as the Regional Water Quality Control Board (Los Angeles Region). She began to learn about the environmental issues surrounding styrofoam and plastic and started focusing her attention on seeing trash as a pollutant rather than just an aesthetic nuisance, which is the way most people saw it at the time.

Zero Waste Washington: Then, in 2001, Heather moved to Seattle with her family where she has worked for multiple environmental groups, most recently joining Zero Waste Washington in 2016. Zero Waste Washington's mission is to make trash obsolete. She assists the organization in working to pass environmental protection legislation at the state, city, and county levels, in addition to organizing research and community projects.

Policy changes: Dramatic and positive waste-reduction change is happening in Washington State and elsewhere. In 2019, the Washington State Legislature passed five of the nine bills Heather and her organization (with partners) promoted. Which ones passed?

1. **Senate Bill 1543**: This law focuses on developing Washington markets for recyclables. It "creates a recycling development center to research, incentivize, and develop new markets and expand existing markets for recycled commodities and recycling facilities. [It r]equires [the Department of] Ecology to . . . implement a statewide recycling contamination reduction and outreach plan."[294] The law will help limit shipping our recyclables internationally.

For more than twenty years, America and much of the world have been shipping the majority of their recycling to China. Then, in 2018, China mostly shut its doors, saying that much of the recycling content that it had been receiving was contaminated, the workers were exposed to toxic chemicals and dangerous working conditions, and the demand for previously desired raw materials, such as certain kinds of plastics, was dwindling. Then, poorer countries, such as India, Vietnam, Malaysia, and Thailand, took up the slack and accepted some of the recycling China refuses,[295] but now, even those countries

---

294 "Legislative Work." Zero Waste Washington, n.d.
295 Mosbergen, Dominique. "China No Longer Wants Your Trash. Here's Why That's Potentially Disastrous." HuffPost. HuffPost, January 25, 2018.

are closing their doors. Cities in the US are now cutting back on what they are willing to recycle. The solution is to build domestic end markets so that our recyclables start to have local value again.

2. **House Bill 5397**: This law focuses on producer responsibility for plastic packaging. In British Columbia and much of Europe, manufacturers pay for the end-of-life of recyclables. Here, they don't. This law requires the collection of data that will help the legislation in 2021 to address the recycling crisis.

3. **House Bill 1114:** This law focuses on food waste. Heather tells me that, in Washington State, 17 percent (by weight) of what's going into our landfills is food. Half of that is edible food. This law helps Washington State preserve and distribute the edible food that would otherwise go into landfills and efficiently compost inedible or spoiled food waste.

4. **House Bill 1569**: This law steps up compostability standards, meaning products once labeled as compostable or biodegradable or "ecodegradable" must now adhere to a compostability standard. To be labeled compostable, items have to be made of a fiber or wood material, or they must comply with composting material requirements outlined by the American Society of Testing and

Materials (ASTM). Items cannot be falsely labeled as biodegradable. Additionally, compostable cutlery and plastic bags have to be tinted brown or green to easily guide consumers in disposing of them properly.

5. **House Bill 1652**: Washingtonians will be able to take their leftover latex paint to stores and other collection sites so that it can be recycled into new paint. Heather says of the paint recycling plants, "They have a computer, and they have a suite of colors that they are producing, and they can literally get that same blue, that same green, that same white, again and again and again, because the computer can calculate it." This recycled paint is cheaper for consumers (about half the cost of new paint) and is high quality. Oil paint will either be redistributed or disposed of as hazardous waste. Unused paint often sits in garages for years or is simply thrown in the landfill.

HEATHER'S THREE DEFINITIONS OF ZERO WASTE:

1. "The one that I like the best is that you are reducing waste from the beginning, so you don't have waste at the end." For example, if you bring your own bags to the grocery store, you don't create additional and unnecessary waste.

2. "Another definition is that you're basically creating a closed loop, so that something is cradle to cradle, not

cradle to grave." An example of cradle to cradle is a car that is designed for easy repair. Fixing instead of replacing is key.

3. "The last definition of zero waste is the concept that waste from the first provides the nutrients for the second. You have a virtuous cycle, basically. The waste of one becomes the nutrients for another." For example, a cherry tree that drops its leaves and blossoms becomes food for the animals nearby.

Heather's thoughts on zero-wasters: "I'm not one of those people who is living a perfect, zero-waste lifestyle. I admire them, but I am not one of those people. We are not perfect and people should not beat themselves up for not being perfect, but should be working toward improving their own practices."

Individual actions make a difference: "It's pretty amazing how quickly actions taken by individuals in their homes end up being felt at a city level. A great example of that is water conservation. Twenty years ago, there was a huge push for people to save water. Low-flow toilets, low-flow showers, and simply not keeping the water running when brushing your teeth. City officials saw a huge decrease in the amount of water being used."

Hope for the future: Heather has hope for the next generation. She says, "My generation and my parents' generation bought into the idea that you should live better than your parents, and there was this consumer lifestyle. The Millennials and Gen Z are definitely way more aware and interested in being proactive on these issues and seeing contradictions and being willing to make and demand big changes across the board—not just environmental but also social and cultural." Heather says that people are willing to take action and make simple changes in their lives once they learn about these issues. Then, these people can spread the word and encourage older people to do the same.

HEATHER'S ADVICE TO LIVE MORE SUSTAINABLY AND HEALTHILY:

1. Don't automatically throw away unsustainable items: "Use them up, don't just throw them away. You have already made the purchase. You have already had the impact. Just go ahead and finish it up, and don't buy it again."

2. Eat as much organic food as possible: "It is so much better for the environment and for your own health to go organic, but it does cost more. However, you can make a few switches that could be cheaper." She says check out the bulk section of the store. If you buy organic beans

at the bulk section, not only will they be cheaper than their canned counterparts, you can bring your own containers, reducing wasteful packaging, and the food won't be contaminated with BPA, which many canned items contain.

3. Buy used clothes instead of new.

4. Only buy toilet paper made of recycled materials. A lot of toilet paper is made from virgin wood, which is unnecessary. Who Gives a Crap is a recyclable toilet paper company that uses no plastic packaging and donates 50 percent of profits to build toilets for people in need.

5. Bring a water bottle with you everywhere you go, so you don't buy disposable plastic.

6. Bring reusable bags to the store.

7. Look at your transportation, housing choices, and what you are consuming. The largest footprint of an item is its production. If you only need to use an item once, borrow it.

8. Watch the film *Plastic China* to learn more about the global recycling crisis.

* * *

Name: Rob Greenfield

Current location: Orlando, Florida

Known for: YouTube channel, Rob Greenfield

Most viewed video: Simple and Sustainable Living In My 100 Square Foot Tiny House

The problem: Rob thought he lived a relatively environmentally-friendly life, but when he started reading books and watching documentaries, he came to understand how his individual actions wreaked havoc on the environment. He realized that he was part of the problem. America throws away $165 billion worth of food per year, while one in seven Americans is food insecure and relies on food banks.[296] Rob says in his 2016 TedX Talk, "I learned that food waste is one of the most pressing environmental issues of our time. When we waste food, we don't just waste the food, we waste all the land, the water, the fossil fuel that was used to grow that food." America has a major food distribution problem.[297]

The solution: Rob realized he could be part of the solution and still be a professional adventurer, as he had dreamed. He

296 "America's $165 Billion Food-waste Problem." CNBC. July 17, 2015.
297 Greenfield, Rob. "How To End The Food Waste Fiasco | Rob Greenfield | TEDxTeen." YouTube. TedX Talks, February 2, 2016.

started environmentally-focused projects that inspired, educated, and motivated people. In 2013, Rob biked across America to show people that they could live more sustainably. On that trip, he secured 70 percent of his diet from dumpsters he found outside of grocery stores. The food he ate was untouched produce or had just expired. Then, on his second bike trip in 2014, Rob focused on the food-waste problem in America. He traveled to various cities and towns, recruited volunteers on Facebook, and went dumpster-diving. Then, they'd share with the world a picture of all the food they found to illustrate the inefficiency of the American food system.

Current project: On November 11, 2018, Rob began his next project in Orlando, Florida, which he calls Project Food Freedom. For the next year, Rob is exclusively growing and foraging his own food. He is not buying food from grocery stores or restaurants. He lives in a 100 square-foot tiny house, which cost him less than $1,500 to build and uses 99 percent repurposed or second-hand materials. He has a small freezer for frozen food. His pantry stores fermented foods, produce he grew or foraged, such as yams, yucca, sweet potatoes, collards, kale, cabbage, papaya, citrus, grapefruit, lemons, oranges, mangoes, bananas, fresh herbs, garlic, coconuts, and even honey from his four beehives. He also goes fishing for another protein source.

Rob tells me in an interview that he spends forty to sixty hours per week focusing on food-related activities and has eaten 130 to 150 different foods since the project began. Outside of his tiny home, he has an outdoor kitchen where he cooks food in a number of ways. He has a propane camp stove, a HomeBiogas stove (he puts food waste inside and it produces sustainable cooking fuel), a fire pit, and a solar oven. He also collects rainwater, which he purifies through the Berkey water filter. And he teaches gardening classes.

The why: "We live in this globalized, industrialized food system where we don't really understand our food very well—where it comes from, how it gets to us. So, for me, it's this deep exploration of, 'Is it possible in the twenty-first century in a modern society to do this, to grow and forage all of my food?' And that's one of the big motivations. Is it possible and what is it like? The other big part of it is, my mission in life is to wake people up, to inspire people—to do my part to change the systems that are not really working for us. I do these extreme projects to catch people's attention and get them thinking and talking about solutions, talking about alternatives, and inspire people to make little changes."

Most time-intensive food to grow or forage: Beans. It takes a long time to shell and process them.

Favorite food to grow: Sweet potatoes, beets, and potatoes. Rob likes root crops.

Aim of his project: Rob wants people to be more connected to their food. They should ask where food comes from, how it gets to us, how people and other species were treated, and what impact it has on our world. Rob hopes that once people educate themselves about their food, they will change the way they source it. He hopes people get their food from local organic farmers or grow it themselves. He wants people to focus on purchasing unpackaged foods and using the bulk section of the grocery store.

How the environment and health are interconnected: "The thing is, we look at our bodies as separate. But the reality is that our bodies are just part of earth. What we do to the earth, we do to ourselves. Sometimes we see it directly, and other times not. Generally, packaged, processed food is not good for the earth—it's not the most environmentally-friendly option, and it's generally not the best option for us, either."

* * *

Name: Shelbi Orme

Current location: Austin, Texas

Known for: YouTube channel, Shelbizleee

Most viewed video: Ulta Dumpster Diving Haul/$2,000+ of Product

The problem: Shelbi majored in environmental science in college and has always been passionate about the earth. She realized how many people lead unsustainable lives, including herself. We aren't trained to question how products are made and how they get to us.

The solution: Shelbi started a YouTube channel and began making sustainability-focused content in July 2017. Her tagline is, "You cannot do all the good that the world needs, but the world needs all the good that you can do." Shelbi shares on YouTube how to live a more eco-friendly lifestyle, whether that involves having a more sustainable period, shopping at well-known grocery stores using the least amount of packaging possible, or traveling in less wasteful ways.

Daily lower waste swaps she loves: While you can make your own skincare products, she loves a brand called Osea, which is a sustainable company that packages its products in glass. She also loves using her bamboo toothbrush, stasher bags (reusable silicone bags instead of Ziploc), a reusable container called a Tiffin for to-go food, and her Pela compostable phone case. She also is a fan of reusable utensils and carries them everywhere.

Her advice: "Do everything you can, but you don't live in a perfect system, so you can't do everything. Zero waste is more of a journey than a destination."

Shelbi's tips for switching to a less wasteful lifestyle:

1. Do a trash audit. Look through your trash bag and evaluate what you're throwing away. Stop your biggest contributors. Check out Shelbi's zero waste transformation series on YouTube for inspiration.

2. Avoid the big four: plastic water bottles, plastic straws, plastic cups, and plastic bags. Use the reusable versions instead.

3. Educate yourself. She recommends the book *101 Ways to Go Zero Waste* by Kathyrn Kellogg.

4. Switch to the search engine Ecosia. The profits from this search engine go toward planting trees—71 million as of October 2019.

5. Calculate your water footprint online. Much of what we consume is virtual water, meaning we don't see the water used, but it was used to produce the items we purchase. She also recommends calculating your carbon footprint online, as well.

6. Check out carbon offsetting programs when planning your travels. These websites calculate the CO_2 emissions caused by your travel and accept your donation to environmental programs that offset those emissions. Shelby likes using www.carbonfund.org.

7. Choose airlines that are more sustainably focused. For example, United Airlines started using biofuels in 2015.[298]

---

298 "United Airlines Makes History with Launch of Regularly Scheduled Flights Using Sustainable Biofuel." United. United Airlines, Inc., January 31, 2019.

SO LONG, FAREWELL, AUF WIEDERSEHEN, GOODBYE

———

I'm going to be honest with you—sometimes I wish I could go back to not knowing all this stuff. Wouldn't it be nice? I wouldn't know that receipt paper has BPA in it. I wouldn't know that make-up can contain asbestos and still remain on store shelves in America. I wouldn't know about the glyphosate in conventional cereals, microplastics filling our oceans, the flame retardants in couches and children's pajamas, the chemicals in the umbilical cords of babies, the heavy metals in fish, the sick people, the exploited workers, the greenwashing, the list goes on. I mean, wouldn't it be easier? More convenient? Wouldn't we be happier? We could just keep living our lives, unaware of all the destruction and suffering and pollution and environmental degradation that is happening all around us.

But, as unsettling and *truly* depressing these realities are, as Amour Vert's founder, Linda Balti, said, "When you know, you can't un-know." The information is out there. If we want to progress as a planet and as a species, we need to consume more intelligently, more sustainably, and more healthfully. We need to do our own research, even though companies should be more transparent. We need to look at a product and understand what we are purchasing—who made it, what it's made of, how we'll use it, and for how long. We need to remember to reduce and reuse before we recycle.

We need to encourage companies to follow triple bottom practices and create sustainable and healthy companies of our own. We need to speak up, both with our voices *and* our dollars. We need to create a culture where disposability and single-use are not the norm.

Abigail Forsyth, the co-founder of KeepCup, a reusable coffee cup company, envisions a world where "convenience culture" is a phenomenon of the past. She wants people of the future to be in absolute disbelief when they look back on people of 2019, asking, "What?! You walked around with a disposable cup?! What for?!" She sees convenience culture as a product of a society that has glorified being too busy and not having enough time for anything, which encourages carelessness and waste. She says, "I think that's definitely changing. Being busy and having no time to do anything is not an aspirational behavior any more, which is significant and will change the way a lot of things happen."

Abigail also envisions an economic capitalist transformation. Currently, we have a "linear economy"—the steps from manufacturing a product to the landfill proceed in a straight line. Manufacturers make products using raw materials, people buy them, use them, and then dispose of them in landfills. Abigail advocates a "circular economy," which utilizes materials already in existence; people use a product, and later, recycle the materials into a new product. She explains, "It's about changing the take, make, waste economy, and keeping products in circulation as long as possible and finding second lives for them. That's changing our entire system."

Jeffrey Hollender, co-founder of Seventh Generation, echoes Abigail's sentiment about a circular economy:

> Are you going to support a company like Patagonia that does many wonderful things, as well as being incredibly politically active, or are you going to buy a coat that is made in China with labor that isn't paid well in very unhealthy

conditions? You have to make the decision. Now the best decision might be not to buy a new coat at all and to go to the Patagonia website and buy a used coat that they have refurbished—that's probably your best choice.

A final word: I don't want the information in this book to paralyze you. I hope you can use the steps at the end of each chapter to start consuming in a healthier and more ethical way. You don't need to totally revamp your life. Just start making little changes, and I promise that steps that once seemed like a hassle will become habitual and easy. All these changes will help you and your family, and they all have an added bonus of helping others and the planet as well. It's a win-win. As I wrote this book and delved deeper into topics, I had many a-ha moments—I hope you did, too. We have only been gifted one planet. It's beautiful and perfect and sacred. Let's protect it.

Thank you for reading this book. Ghandi once said, "We must be the crunchy change we wish to see in the world." Okay, he didn't say "crunchy," but I bet he would have been on board. It's time to get crunchy. It's time for both consumers and businesses to take collective action to create a better world. After all, all the cool kids are doing it.

ACKNOWLEDGMENTS

I have always wanted to write a book. This goal has shown up on my list of New Year's Resolutions since before I can remember. However, this goal kept getting postponed until I received an email in December of 2018 from Drew Dudley, a fellow English major at University of Washington. She was reaching out to students who were interested in writing a book over the subsequent two quarters. The program she described to me was called Creator Institute and was led by a Georgetown professor named Eric Koester.

I cyberstalked Professor Koester, watched his talk on You-Tube, thought the idea was cool, and then convinced myself that writing a book was too scary, daunting, and overwhelming. It would be too much work and take up too much energy. I closed my laptop and tried to mentally close the tab that was this opportunity. But, I couldn't stop thinking about it. I had to take it. Fast forward to December, 2019, and *It's Easy Being Green: How Conscious Consumers and Ecopreneurs Can Save the World* is now in your hands. I couldn't have made my book dream a reality had it not been for the many talented and kind people who helped me, encouraged me, and motivated me to keep going.

First and foremost, I'd like to thank Professor Eric Koester. Thank you for running this program, guiding me, and showing the world that anyone can write a book at any age regardless of experience. I'd also like to thank Drew Dudley, our UW chapter leader, and Brian Bies, the Head of Publishing at New Degree Press. Thanks also to my developmental editor, Shelby Hogan, my marketing editor, Elina Oliferovskiy, and my copy editor, Kayla LeFevre. Thank you all for helping bring my vision to life. I appreciate each of you.

I'd also like to thank my parents, Steve and Nena, for supporting me. Thank you for your encouragement, love, and enthusiasm. Thank you for hearing my ideas about this book, taking the time to read my manuscript, and giving me invaluable feedback. I know you made this book more cohesive and digestible, and I am grateful. Thank you for instilling within me the idea that I can always strive for more and make an impact, regardless of my circumstances.

I'd also like to thank Benedikt H. Thank you for your support, speedy feedback, and exclamation marks. I know you made this book better. I am grateful and appreciate you. Vielen Dank.

Thank you also to Lindsay Christensen for giving me feedback. Much appreciated.

Also, thank you to all the people who let me interview them. Thanks to each of you for taking the time out of your schedule to chat with me and share your dreams, stories, and insights. I appreciate each of you. You all are doing such important work. Thank you for helping create a better world.

Lastly, I'd like to thank all my Indiegogo supporters (in alphabetical order) for backing my dream. Thank you to:

Donald Baker

Spencer & Sam Baker

Kiley Beck

Sorcha Connor-Boyle

Keziah Serwaa Buabeng

Steve Butler

Jenny Buttaccio

K Cervenka

Aarti Chandorkar

Lindsay Christensen

Kathleen Conner

Shannon Donegan

Jen Dorée

Laura Ehlers

Thomas Empson

Suzy Ettinger

Tobyanna Everhart

Victoria Faling

Anna Fleetwood

Abigail Forsyth

Joan Forsythe

Judy & Arnold Fox

Michelle Fox

Rachel Funk

Tierra Garcia

Laura Gibbons

Sheila & Peter Gluck

Jenn Godwin

Michele Goldman

Benedikt H.

Yvonne Hall

Lianne Hart-Thompson

Janell Hartman

Elizabeth Heile

Megan Horowitz

Chris Kautsky

David Keyes

Samia Khudari

David Kirdahy

Kodjovi Klikan

Eric Koester

Josh LaBelle

Liz Leshin

Hillary Brook Levy

Jenna Levy

Kami Lingren

Courtney Loftus

Jamie Lynn

Pamela Marcus

Andrea McCoy

Shakirra Meghjee

Ashley Mersereau

Paul Messinger

Susan Messinger

Clara Midgley

Harvey Motulsky

Emily Nichols

Jane & Jerry Ninteman

Lisa Norton

Martha Padilla

Charlie Peirson

Lindsay Peltin

Saree & Bill Peltin

Sherwin & Julie Peltin

Barry Perzow

Francie Petracca

Rachael & Rudy Prather

Kaeley Pruitt-Hamm

Jennifer Reibman

Emily Rosner

Amy Sanchez

Christina Serkowski

Lynn Shelton

Corinne Sternberg

Ann Margaret Stompro

Alaina Swick

Heather Trim

Carol Vare

Jeremy Vassallo

Marie Whalen

Gosia Wolfe

Dan Wuthrich

Cynthia Young

Laura Caldwell Zaugg

Kristi & Tim Zimmer

WORKS REFERENCED

———

INTRODUCTION

EBC. "Crunchy." Urban Dictionary. Urban Dictionary, June 14, 2007. https://www.urban-dictionary.com/define.php?term=crunchy.

Group, Edward. "5 Dangerous Chemicals in Sunscreen." Global Healing Center. Global Healing Center, October 21, 2015. https://www.globalhealingcenter.com/natural-health/5-dangerous-chemicals-in-sunscreen/.

Kenton, Will. "Triple Bottom Line (TBL)." Investopedia. Investopedia, May 3, 2019. https://www.investopedia.com/terms/t/triple-bottom-line.asp.

Lappé, Anna. "Anna Lappé > Quotes > Quotable Quote." Goodreads. Goodreads, n.d. https://www.goodreads.com/quotes/587323-every-time-you-spend-money-you-re-casting-a-vote-for.

Matta, Murali K, Robbert Zusterzeel, Nageswara R Pilli, Vikram Patel, Donna A Volpe, Jeffry Florian, Luke Oh, et al. "Effect of Sunscreen Application Under Maximal Use Conditions on Plasma Concentration of Sunscreen Active Ingredients: A Randomized Clinical Trial." JAMA. American Medical Association, June 4, 2019. https://www.ncbi.nlm.nih.gov/pubmed/31058986/.

Mercola, Dr. "More Than Half of Americans Have Chronic Illnesses." Mercola.com. Mercola, November 30, 2016. https://articles.mercola.com/sites/articles/archive/2016/11/30/expensive-us-health-care.aspx.

Morris, Zoë Slote, Steven Wooding, and Jonathan Grant. "The Answer Is 17 Years, What Is the Question: Understanding Time Lags in Translational Research." Journal of the Royal Society of Medicine. Royal Society of Medicine Press, December 2011. https://www.ncbi.nlm.nih.gov/pmc/articles/PMC3241518/.

Scheer, Roddy, and Doug Moss. "Why Are Trace Chemicals Showing Up in Umbilical Cord Blood?" Scientific American. Scientific American, September 1, 2012. https://www.scientificamerican.com/article/chemicals-umbilical-cord-blood/.

"Toxic Chemicals Released by Industries This Year, Tons." Worldometers. Worldometers, n.d. https://www.worldometers.info/view/toxchem/.

CHAPTER 1

Brandt, Allan M. "Inventing Conflicts of Interest: A History of Tobacco Industry Tactics." American Journal of Public Health. PMC, January 2012. https://www.ncbi.nlm.nih.gov/pmc/articles/PMC3490543/.

Callahan, Patricia, and Sam Roe. "Flame Retardants: A Dangerous Lie." The Telegraph. The Telegraph, May 26, 2012. https://www.macon.com/news/local/article30106617.html.

"Dental Amalgam Mercury Fillings and Danger to Human Health." IAOMT. The International Academy of Oral Medicine & Toxicology, 2016. https://iaomt.org/resources/dental-mercury-facts/amalgam-fillings-danger-human-health/.

Duckworth, Lorna. "Scientists Link Abraham Lincoln's Fits of Rage to Mercury Poisoning." The Independent. Independent Digital News and Media, July 18, 2001. https://www.independent.co.uk/news/science/scientists-link-abraham-lincolns-fits-of-rage-to-mercury-poisoning-9223021.html.

"Firefighter Calls for Action on Toxic Flame Retardant Chemicals." NRDC. June 11, 2019. Accessed June 23, 2019. https://www.nrdc.org/stories/firefighter-calls-action-toxic-flame-retardant-chemicals.

King, Delaney. "You Can Find Hundreds (!) of Couches Without Toxic Flame Retardants." EWG. EWG, August 31, 2016. https://www.ewg.org/childrenshealth/20888/you-can-find-hundreds-couches-without-toxic-flame-retardants.

Little, Becky. "When Cigarette Companies Used Doctors to Push Smoking." History.com. A&E Television Networks, September 13, 2018. https://www.history.com/news/cigarette-ads-doctors-smoking-endorsement.

McKay, Jim. "Firefighters Turn to Chemical Detox Saunas to Thwart the Cancer Threat." Government Technology State & Local Articles - E.Republic. April 3, 2018. Accessed June 23, 2019. https://www.govtech.com/em/disaster/Firefighters-Turn-to-Chemical-Detox-Saunas-to-Thwart-the-Cancer-Threat.html.

Mendes, Elizabeth. "The Study That Helped Spur the U.S. Stop-Smoking Movement." American Cancer Society. American Cancer Society, January 9, 2014. https://www.cancer.org/latest-news/the-study-that-helped-spur-the-us-stop-smoking-movement.html.

Michon, Kathleen. "Tobacco Litigation: History & Recent Developments." NOLO. NOLO, 2019. https://www.nolo.com/legal-encyclopedia/tobacco-litigation-history-and-development-32202.html.

Miriam Varkey, Indu, Rajmohan Shetty, and Amitha Hegde. "Mercury Exposure Levels in Children with Dental Amalgam Fillings." International Journal of Clinical Pediatric Dentistry. Jaypee Brothers Medical Publishers, February 9, 2015. https://www.ncbi.nlm.nih.gov/pmc/articles/PMC4335109/.

More Doctors Smoke Camels Than Any Other Cigarette. YouTube. YouTube, 2006. https://www.youtube.com/watch?v=gCMzjJjuxQI&frags=pl,wn.

"About Dental Amalgam Fillings." U.S. Food and Drug Administration. FDA, n.d. https://www.fda.gov/medical-devices/dental-amalgam/about-dental-amalgam-fillings#risks.

"Precautionary Principle." ScienceDirect. Elsevier B.V. , 2018. https://www.sciencedirect.com/topics/earth-and-planetary-sciences/precautionary-principle.

"Toxic Flame Retardants." Safer States. Accessed June 10, 2019. http://www.saferstates.com/toxic-chemicals/toxic-flame-retardants.

CHAPTER 2

Behari, Jitendra, and Paulraj Rajamani. "Electromagnetic Field Exposure Effects (ELF-EMF and RFR) on Fertility and Reproduction." Semantic Scholar. Semantic Scholar, November 2012. https://pdfs.semanticscholar.org/d1df/2ab3f1daf6ef7cd-8be364489b64919919d11.pdf.

Dasdag, Suleyman, and Mehmet Zulkuf Akdag. "The Link between Radiofrequencies Emitted from Wireless Technologies and Oxidative Stress." Journal of Chemical Neuroanatomy. ScienceDirect, September 12, 2015. https://www.sciencedirect.com/science/article/pii/S0891061815000691.

Degnan, Peter M., Scott A. McLaren, and Michael T. Tennant. " Telecommunications Act of 1996: 704 of the Act and Protections Afforded the Telecommunications Provider in the Facilities Sitting Context, The." Michigan Law. Michigan Law, 1997. https://repository.law.umich.edu/cgi/viewcontent.cgi?referer=&httpsredir=1&article=1167&context=mttlr.

Dovey, Dana. "A Switch to 5G May Be Bad for the Environment." Newsweek. Newsweek, August 27, 2018. https://www.newsweek.com/migratory-birds-bee-navigation-5g-technology-electromagnetic-radiation-934830.

"Electrohypersensitivity Overview." Physicians for Safe Technology. Physicians for Safe Technology, September 26, 2017. https://mdsafetech.org/problems/electro-sensitivity/electrosensitivity-history/.

"France: New National Law Bans WIFI in Nursery School!" Environmental Health Trust. Environmental Health Trust, October 28, 2015. https://ehtrust.org/france-new-national-law-bans-wifi-nursery-school.

Generation Zapped, 2017.

Gross, Terry. "After Dump, What Happens To Electronic Waste?" NPR. NPR, December 21, 2010. https://www.npr.org/2010/12/21/132204954/after-dump-what-happens-to-electronic-waste.

Havas, Magda. "Radiation from Wireless Technology Affects the Blood, the Heart, and the Autonomic Nervous system1)." Reviews on Environmental Health. De Gruyter, November 5, 2013. https://www.degruyter.com/view/j/reveh.2013.28.issue-2-3/reveh-2013-0004/reveh-2013-0004.xml.

Keller-Byrne, Jane E., and Farhang Farhang Akbar-Khanzadeh. "Potential Emotional and Cognitive Disorders Associated with Exposure to EMFs." Sage Journals. Sage Journals, February 1997. https://journals.sagepub.com/doi/pdf/10.1177/216507999704500205.

Lai, Henry. "Neurological Effects of Radiofrequency Electromagnetic Radiation Relating to Wireless Communication Technology." The EMR Policy Institute . University of Washington, 1997. http://www.emrpolicy.org/science/forum/laibrussels.pdf.

Pall, Martin L. "Microwave Frequency Electromagnetic Fields (EMFs) Produce Widespread Neuropsychiatric Effects Including Depression." Journal of Chemical Neuroanatomy. ScienceDirect, August 21, 2015. https://www.sciencedirect.com/science/article/pii/S0891061815000599.

Pineault, Nick. *The Non-Tinfoil Guide to EMFs: How To Fix Our Stupid Use of Technology*, N&G Média Inc., 2017.

Rambaldi, Massimiliano. "L'elettrosmog Fa Chiudere Il Vecchio Parco Giochi - La Stampa." La Stampa. La Stampa, April 26, 2019. https://www.lastampa.it/torino/2019/04/26/news/l-elettrosmog-fa-chiudere-il-vecchio-parco-giochi-1.33697683.

Rea, William J, Yaqin Pan, Ervin J. Yenyves, Iehiko Sujisawa, Hideo Suyama, Nasrola Samadi, and Gerald H. Ross. "Electromagnetic Field Sensitivity Case Study Evaluation." Journal of Bioelectricity, 1991. https://pdfs.semanticscholar.org/458d/36a2b536ca4c22f-5c49e7a5637e28310e5d8.pdf.

Redmayne, Mary, Euan Smith, and Michael J. Abramson. "The Relationship between Adolescents' Well-Being and Their Wireless Phone Use: a Cross-Sectional Study." Environmental Health. BioMed Central, October 22, 2013. https://ehjournal.biomedcentral.com/articles/10.1186/1476-069X-12-90.

West, John G, Nimmi S Kapoor, Shu-Yuan Liao, June W Chen, Lisa Bailey, and Robert A Nagourney. "Multifocal Breast Cancer in Young Women with Prolonged Contact Between Their Breasts and Their Cellular Phones." Case Reports in Medicine. Hindawi Publishing Corporation, 2013. https://www.ncbi.nlm.nih.gov/pmc/articles/PMC3789302/.

Yildirim, Mehmet Erol, Mehmet Kaynar, Huseyin Badem, Mucahıt Cavis, Omer Faruk Karatas, and Ersın Cimentepe. "What Is Harmful for Male Fertility: Cell Phone or the Wireless Internet?" The Kaohsiung Journal of Medical Sciences. ScienceDirect, July 26, 2015. https://www.sciencedirect.com/science/article/pii/S1607551X1500162X.

CHAPTER 3

"About the FTC." Federal Trade Commission. FTC, n.d. https://www.ftc.gov/about-ftc.

"Careers at the FTC." Federal Trade Commission, n.d. https://www.ftc.gov/about-ftc/careers-ftc.

Chen, Caroline. "FDA Repays Industry by Rushing Risky Drugs to Market." ProPublica, March 9, 2019. https://www.propublica.org/article/fda-repays-industry-by-rushing-risky-drugs-to-market.

"Children's Sleepwear Regulations." CPSC.gov. CPSC, April 10, 2019. https://www.cpsc.gov/Business--Manufacturing/Business-Education/Business-Guidance/Childrens-Sleepwear-Regulations/.

Cook, Ken. "Ken Cook on EWG's 20th Anniversary." YouTube. YouTube, October 28, 2013. https://www.youtube.com/watch?v=HbBoHVS6h64.

"Cosmetics Safety Q&A: Prohibited Ingredients." U.S. Food and Drug Administration. FDA, n.d. https://www.fda.gov/cosmetics/resources-consumers-cosmetics/cosmetics-safety-qa-prohibited-ingredients.

"DETAIL OF FULL-TIME EQUIVALENT EMPLOYMENT (FTE)." FDA. FDA, 2018. https://www.fda.gov/media/106372/download.

Frack, Lisa, and Becky Sutton. "3,163 Ingredients Hide Behind the Word 'Fragrance.'" EWG. EWG, February 2, 2010. https://www.ewg.org/enviroblog/2010/02/3163-ingredients-hide-behind-word-fragrance.

"Fragrances in Cosmetics." U.S. Food and Drug Administration. FDA, n.d. https://www.fda.gov/cosmetics/cosmetic-ingredients/fragrances-cosmetics.

"Frequently Asked Questions and Answers (FAQs)." CPSC.gov. CPSC, n.d. https://www.cpsc.gov/About-CPSC/Contact-Information.

"Glyphosate." Wikipedia. Wikimedia Foundation, June 7, 2019. https://en.wikipedia.org/wiki/Glyphosate.

"H&M Recalls Children's Pajamas Due to Violation of Federal Flammability Standard." U.S. Consumer Product Safety Commission, July 25, 2019. https://www.cpsc.gov/Recalls/2019/H&M-Recalls-Childrens-Pajamas-Due-to-Violation-of-Federal-Flammability-Standard.

"Identity Theft." Federal Trade Commission. FTC, n.d. https://www.ftc.gov/news-events/media-resources/identity-theft.

"Mergers and Competition." Federal Trade Commission. FTC, n.d. https://www.ftc.gov/news-events/media-resources/mergers-and-competition.

"Mobile Technology Issues." Federal Trade Commission, n.d. https://www.ftc.gov/news-events/media-resources/mobile-technology.

"Monsanto Papers." U.S. Right to Know. USRTK, n.d. https://usrtk.org/monsanto-papers/.

Milman, Oliver. "US Cosmetics Are Full of Chemicals Banned by Europe – Why?" The Guardian. Guardian News and Media, May 22, 2019. https://www.theguardian.com/us-news/2019/may/22/chemicals-in-cosmetics-us-restricted-eu.

"Non-Toxic Diapers: Safer Disposable Diapers for Babies." The Gentle Nursery. The Gentle Nursery , n.d. https://www.gentlenursery.com/baby-care/non-toxic-diapers/.

"Questions and Answers on Glyphosate." U.S. Food and Drug Administration. FDA, n.d. https://www.fda.gov/food/pesticides/questions-and-answers-glyphosate.

Saltzburg, Tara. "Snug Fit Pajamas: A Guide to Flame Retardants and the Children's Sleepwear Regulations." Westyn Baby, February 14, 2019. https://www.westynbaby.com/blogs/westyn-baby/snug-fit-pajamas-guide-to-flame-retardants-and-childrens-sleepwear-regulations.

Schrock, Monica. "WTF Is In Fragrance and Is It Harmful!?" Non Toxic Revolution. Non Toxic Revolution, May 30, 2017. https://www.nontoxicrevolution.org/blog/wtf-fragrance.

"Small Businesses & Homemade Cosmetics: Fact Sheet." U.S. Food and Drug Administration. FDA, n.d. https://www.fda.gov/cosmetics/resources-industry-cosmetics/small-businesses-homemade-cosmetics-fact-sheet#2.

"Super Jumper Recalls Trampolines Due to Fall and Injury Hazards." U.S. Consumer Product Safety Commission, August 1, 2019. https://www.cpsc.gov/Recalls/2019/Super-Jumper-Recalls-Trampolines-Due-to-Fall-and-Injury-Hazards.

"The Do Not Call Registry." Federal Trade Commission. FTC, n.d. https://www.ftc.gov/news-events/media-resources/do-not-call-registry.

"Truth In Advertising." Federal Trade Commission. FTC, n.d. https://www.ftc.gov/news-events/media-resources/truth-advertising.

Tse, Lily. "A MESSAGE FROM THE FOUNDER." Think Dirty. Think Dirty, n.d. https://www.thinkdirtyapp.com/about/.

"What We Do." U.S. Food and Drug Administration. FDA, n.d. https://www.fda.gov/about-fda/what-we-do.

"When and Why Was FDA Formed?" U.S. Food and Drug Administration. FDA, n.d. https://www.fda.gov/about-fda/fda-basics/when-and-why-was-fda-formed.

"Wintergreen Essential Oil Recalled by Epic Business Services Due to Failure to Meet Child Resistant Closure Requirement; Risk of Poisoning (Recall Alert)." U.S. Consumer Product Safety Commission, July 16, 2019.

https://www.cpsc.gov/Recalls/2019/Wintergreen-Essential-Oil-Recalled-by-Epic-Business-Services-Due-to-Failure-to-Meet-Child-Resistant-Closure-Requirement-Risk-of-Poisoning-Recall-Alert.

CHAPTER 4

"How To Get A Certified Organic Product Label (in the USA)." Quick-Label Blog, February 20, 2019. https://blog.quicklabel.com/2010/10/how-to-get-a-certified-organic-product-label-in-the-usa/.

"Methylisothiazolinone and Methylchloroisothiazolinone." Campaign for Safe Cosmetics, n.d. http://www.safecosmetics.org/get-the-facts/chemicals-of-concern/methylisothiazolinone/.

"Organic Labeling Standards." Organic Labeling Standards | Agricultural Marketing Service. USDA, n.d. https://www.ams.usda.gov/grades-standards/organic-labeling-standards.

Saltzburg, Tara. "5 Things To Look For When Buying Children's Sleepwear." Westyn Baby, n.d. https://www.westynbaby.com/blogs/westyn-baby/5-things-to-look-for-when-shopping-for-childrens-sleepwear.

Temkin, Alexis. "Breakfast With a Dose of Roundup?" EWG. EWG, August 15, 2018.

https://www.ewg.org/childrenshealth/glyphosateincereal/#.W3RbFJNKjUI.

CHAPTER 5

"About Proposition 65." OEHHA. OEHHA, n.d. https://oehha.ca.gov/proposition-65/about-proposition-65.

"Are Your Medications Causing Nutrient Deficiency?" Harvard Health. Harvard University, August 2016. https://www.health.harvard.edu/staying-healthy/are-your-medications-causing-nutrient-deficiency.

Marshall, John. "Why You See Such Weird Drug Commercials on TV All the Time." Thrillist. Thrillist, March 23, 2016. https://www.thrillist.com/health/nation/why-are-prescription-drug-advertisements-legal-in-america.

Preidt, Robert. "Americans Taking More Prescription Drugs Than Ever: Survey." Consumer HealthDay. HealthDay, August 3, 2017. https://consumer.healthday.com/general-health-information-16/prescription-drug-news-551/americans-taking-more-prescription-drugs-than-ever-survey-725208.html.

Qato, Dima Mazen, Shannon Zenk, Jocelyn Wilder, Rachel Harrington, Darrell Gaskin, and G. Caleb Alexander. "The Availability of Pharmacies in the United States: 2007–2015." PLOS ONE. Public Library of Science, August 16, 2017. https://journals.plos.org/plosone/article?id=10.1371/journal.pone.0183172.

"Summary of Studies on Food Dyes." Center for Science in the Public Interest. Center for Science in the Public Interest, n.d. https://cspinet.org/sites/default/files/attachment/dyes-problem-table.pdf.

"THE DIETARY SUPPLEMENT CONSUMER: 2015 CRN CONSUMER SURVEY ON DIETARY SUPPLEMENTS." CRN The Science Behind Supplements. CRN, n.d. http://www.crnusa.org/CRN-consumersurvey-archives/2015/.

Tompkins, Tiffany. "Supplement Law Makers Brace for Federal GMO Labeling Law." Compass Natural Marketing. Compass Natural Marketing, October 8, 2016. https://www.compassnaturalmarketing.com/compass-natural-new-directions-for-green-business/2016/10/8/supplement-law-makers-brace-for-federal-gmo-labeling-law.

"Top 10 Magnesium Supplements." Labdoor. Labdoor, n.d. https://labdoor.com/rankings/magnesium?fbclid=IwAR1P8JMxqWGeYrSMtQaG8tzJ_K2JsfCsbk2fD2j_VCacYd4A4G-MZk_FPKEc.

"U.S. Revenue Vitamins & Supplements Manufacturing 2019." Statista. Statista, n.d. https://www.statista.com/statistics/235801/retail-sales-of-vitamins-and-nutritional-supplements-in-the-us/.

Vien, Anya. "Shock Finding: Top Pharma-Brands of Vitamins Contain Aspartame, GMOs, and Hazardous Chemicals." Anya Vien, May 4, 2019. https://anyavien.com/top-pharma-brands-of-vitamins-contain-aspartame/?fbclid=IwAR2C31nZSV-xGa0qw1s-jEVXr5zVNu3FhlQVRMXuLE7LlzrtyhGTVrLjr2AI.

"Worldwide Pharmaceutical Sales by Region 2016-2018 | Statistic." Statista. Statista, n.d. https://www.statista.com/statistics/272181/world-pharmaceutical-sales-by-region/.

Yigzaw, Erika. "5 Dangerous Ingredients in Your Vitamins and Dietary Supplements: Achs.edu." Accredited Online Holistic Health College, December 2, 2016. http://info.achs.edu/blog/dangerous-supplement-ingredients.

CHAPTER 6

"Cosmetics Safety Q&A: Prohibited Ingredients." U.S. Food and Drug Administration. FDA, n.d. https://www.fda.gov/cosmetics/resources-consumers-cosmetics/cosmetics-safety-qa-prohibited-ingredients.

"Exposures Add up – Survey Results | Skin Deep® Cosmetics Database." EWG's Skin Deep. EWG, n.d. https://www.ewg.org/skindeep/2004/06/15/exposures-add-up-survey-results/.

"Why Beautycounter Bans More Ingredients Than The U.S." Beautycounter. Beautycounter, August 1, 2018. https://blog.beautycounter.com/why-beautycounter-bans-more-ingredients-than-the-u-s/.

Yigzaw, Erika. "5 Dangerous Ingredients in Your Vitamins and Dietary Supplements: Achs.edu." American College of Healthcare Sciences. American College of Healthcare Sciences, December 2, 2016. http://info.achs.edu/blog/dangerous-supplement-ingredients.

CHAPTER 7

Bienkowski, Brian. "BPA May Prompt More Fat in the Human Body." Scientific American. May 29, 2015. https://www.scientificamerican.com/article/bpa-may-prompt-more-fat-in-the-human-body/.

Bittner, George D., Chun Z. Yang, and Matthew A. Stoner. "Estrogenic Chemicals Often Leach from BPA-free Plastic Products That Are Replacements for BPA-containing Polycarbonate Products." Environmental Health : A Global Access Science Source. May 28, 2014. https://www.ncbi.nlm.nih.gov/pmc/articles/PMC4063249/.

"Endocrine Disruptors." National Institute of Environmental Health Sciences. https://www.niehs.nih.gov/health/topics/agents/endocrine/index.cfm.

"Fact Sheet: Plastics in the Ocean." Earth Day Network. April 05, 2018. https://www.earth-day.org/2018/04/05/fact-sheet-plastics-in-the-ocean/.

Fox, Kieran D., Garth A. Covernton, Hailey L. Davies, John F. Dower, Francis Juanes, and Sarah E. Dudas. "Human Consumption of Microplastics." Environmental Science & Technology. June 5, 2019. https://pubs.acs.org/doi/abs/10.1021/acs.est.9b01517.

Geller, Samara. "BPA in Canned Food." EWG. June 3, 2013. https://www.ewg.org/research/bpa-canned-food.

Genuis, Stephen J., Sanjay Beesoon, Detlef Birkholz, and Rebecca A. Lobo. "Human Excretion of Bisphenol A: Blood, Urine, and Sweat (BUS) Study." Journal of Environmental and Public Health. Hindawi, 2012. https://www.hindawi.com/journals/jeph/2012/185731/abs/.

Hamilton, Jon. "Beyond BPA: Court Battle Reveals A Shift In Debate Over Plastic Safety." NPR. February 16, 2015. https://www.npr.org/sections/health-shots/2015/02/16/385747786/beyond-bpa-court-battle-reveals-a-shift-in-debate-over-plastic-safety.

Hamilton, Jon. "Study: Most Plastics Leach Hormone-Like Chemicals." NPR. NPR, March 2, 2011. https://www.npr.org/2011/03/02/134196209/study-most-plastics-leach-hormone-like-chemicals.

"Independent Laboratory Testing Results of Popular Gravity Water Filters." WaterFilterLabs.com. http://waterfilterlabs.com/.

Marques-Pinto, André, and Davide Carvalho. "Human Infertility: Are Endocrine Disruptors to Blame? in: Endocrine Connections Volume 2 Issue 3 (2013)." Endocrine Connections. BioScientifica, October 19, 2018. https://ec.bioscientifica.com/view/journals/ec/2/3/R15.xml?fbclid=IwARojIhHQsanFx6cfHisHNCfgxNjGKNwrITAKMHhddE2CFIy3H2xL6h9E678.

Pawlowski, A. "Left Your Bottled Water in a Hot Car? Drink It with Caution, Some Experts Say." Today. July 06, 2018. https://www.today.com/health/bottled-water-hot-plastic-may-leach-chemicals-some-experts-say-t132687.

"PCC Natural Markets Is First Grocer in the Nation to Offer New, Safer Customer Receipts at All of Its Locations." PCC Community Markets. 2014. https://www.pccmarkets.com/news/2014/0930-safer-receipts/.

Shultz, David. "Americans Eat More than 50,000 Tiny Pieces of Plastic Every Year." Science. June 05, 2019. https://www.sciencemag.org/news/2019/06/americans-eat-more-50000-tiny-pieces-plastic-every-year?fbclid=IwAR2nVQOVaVqHRCK-53pypvQAtTJMouVjwQolWhv8GJhwkYXl4VjajJrzsONIa.

"The Global Market For Plastic Products Will Be Worth $1.2 Trillion By 2020." Resource Recycling Inc. https://resource-recycling.com/resourcerecycling/wp-content/uploads/2017/12/TBRC-Plastic-Products.pdf.

Weise, Elizabeth. "Turns out There's More Plastic Pollution in the Deep Ocean than the Great Pacific Garbage Patch." USA Today. June 09, 2019. https://www.usatoday.com/story/news/nation/2019/06/06/forget-great-pacific-garbage-patch-theres-more-plastic-deep-sea/1349571001/.

CHAPTER 8

Association, Press. "Contraceptive Pill 'Can Lead Women to Choose Wrong Partner'." The Guardian. Guardian News and Media, August 13, 2008. https://www.theguardian.com/science/2008/aug/13/medicalresearch.medicaladvicefortravellers.

Brighten, Jolene. *Beyond the Pill: A 30-Day Program to Balance Your Hormones, Reclaim Your Body, and Reverse the Dangerous Side Effects of the Birth Control Pill.* Harper One, 2019.

Das, Reenita. "Women's Healthcare Comes Out Of The Shadows: Femtech Shows The Way To Billion-Dollar Opportunities." Forbes. April 12, 2018. https://www.forbes.com/sites/reenitadas/2018/04/12/womens-healthcare-comes-out-of-the-shadows-femtech-shows-the-way-to-billion-dollar-opportunities/#2775bb46159e.

"How Effective Is Contraception at Preventing Pregnancy?" *NHS Choices*, NHS, 30 June 2017, https://www.nhs.uk/conditions/contraception/how-effective-contraception/.

"How to Chart Your Cycle to Know When You Can Get Pregnant." WebMD. WebMD, n.d. https://www.webmd.com/baby/charting-your-fertility-cycle.

"Natural, Self-Determined Fertility Tracking with Daysy." Daysy. Daysy, March 12, 2019. https://usa.daysy.me/media/filer_public/51/00/51006827-4434-457a-9b28-abee86f4f5d3/factsheet_-_daysy.pdf.

Popsci. "New Study: Fertile Strippers Make More Money." Popular Science. Popular Science, October 5, 2007. https://www.popsci.com/article/2007-10/new-study-fertile-strippers-make-more-money/.

CHAPTER 9

"Abortion Worldwide 2017." Guttmacher Institute. 2018. https://www.guttmacher.org/sites/default/files/report_pdf/abortion-worldwide-2017.pdf.

Acciardo, Kelli. "4 Harmful Lube Ingredients You Should Avoid At All Costs." Prevention. June 10, 2019. https://www.prevention.com/sex/g20482085/personal-lubricant-ingredients-to-avoid/.

Barnes, Zahra. "6 Lube Ingredients You Might Not Want to Put in Your Vagina." SELF. December 01, 2017. https://www.self.com/story/6-lube-ingredients-to-avoid.

"Child Labour in Plantation." Child Labour in Plantation. April 22, 2010.

http://www.ilo.org/jakarta/areasofwork/WCMS_126206/lang—en/index.htm.

Herbenick, Debby, Vanessa Schick, Stephanie A. Sanders, Michael Reece, and J. Dennis Fortenberry. "Pain Experienced During Vaginal and Anal Intercourse with Other-Sex Partners: Findings from a Nationally Representative Probability Study in the United States." The Journal of Sexual Medicine. February 04, 2015. https://onlinelibrary.wiley.com/doi/abs/10.1111/jsm.12841.

McCammon, Sarah. "U.S. Abortion Rate Falls To Lowest Level Since Roe v. Wade." NPR. January 17, 2017. https://www.npr.org/sections/thetwo-way/2017/01/17/509734620/u-s-abortion-rate-falls-to-lowest-level-since-roe-v-wade.

Nicole, Wendee. "A Question for Women's Health: Chemicals in Feminine Hygiene Products and Personal Lubricants." National Institute of Environmental Health Sciences. March 1, 2014. https://ehp.niehs.nih.gov/doi/10.1289/ehp.122-a70.

Normile, Dennis. "The Tires on Your Car Threaten Asian Biodiversity." Science. December 10, 2017. https://www.sciencemag.org/news/2015/04/tires-your-car-threaten-asian-biodiversity

Seppanen, Jahla. "Woah, Should You Be Using Organic Condoms?" Shape. https://www.shape.com/lifestyle/sex-and-love/organic-condoms.

"Sex and HIV Education." Guttmacher Institute. June 03, 2019. https://www.guttmacher.org/state-policy/explore/sex-and-hiv-education.

"Some e-Cigarette Ingredients Are Surprisingly More Toxic than Others." Medical Xpress. Medical Xpress, March 27, 2018. https://medicalxpress.com/news/2018-03-e-cigarette-ingredients-surprisingly-toxic.html.

Stanger-Hall, Kathrin F., and David W. Hall. "Abstinence-Only Education and Teen Pregnancy Rates: Why We Need Comprehensive Sex Education in the U.S." PLOS ONE. October 14, 2011. https://journals.plos.org/plosone/article?id=10.1371/journal.pone.0024658.

"Studies About Why Men And Women Use Lubricants During Sex." ScienceDaily. November 09, 2009. https://www.sciencedaily.com/releases/2009/11/091109090431.htm.

Suzdaltsev, Jules. "What Students in Europe Learn That Americans Don't." Vice. March 16, 2016. https://www.vice.com/en_us/article/3b48d3/what-students-in-europe-learn-that-americans-dont.

"We Can Learn a Lot from the Netherlands' Approach to Sex-ed." Facebook Watch. https://www.facebook.com/watch/?v=1823611481277138.

"What Are Petrochemicals? - The Honest Company Blog." The Honest Company Blog. January 03, 2019. https://blog.honest.com/what-are-petrochemicals/.

"Women's Health Care Physicians." ACOG. https://www.acog.org/Patients/FAQs/When-Sex-Is-Painful#how.

CHAPTER 10

Chavez-MacGregor, Mariana, Carla H. van Gils, Yvonne T. van der Schouw, Evelyn Monninkhof, Paulus A.H. van Noord, and Petra H.M. Peeters. "Lifetime Cumulative Number of Menstrual Cycles and Serum Sex Hormone Levels in Postmenopausal Women." Breast Cancer Research and Treatment. PMC, March 2008. https://www.ncbi.nlm.nih.gov/pmc/articles/PMC2244694/.

Elsworthy, Emma. "More than 137,700 Girls in UK Missed School in the Last Year Because They Couldn't Afford Sanitary Products." The Independent. March 7, 2018. https://www.independent.co.uk/news/international-womens-day-period-girls-missed-school-uk-sanitary-products-menstruation-a8244396.html.

Graham, Karen. "Argentina Study: 85% of Tampons Contaminated with Glyphosate." Digital Journal. October 27, 2015. http://www.digitaljournal.com/life/health/85-percent-of-tampons-found-to-contain-glyphosate/article/447733.

Grahn, Judy. "Chapter 1." Chapter 1: Blood, Bread, and Roses: How Menstruation Created the World. http://bailiwick.lib.uiowa.edu/wstudies/grahn/chapt01.htm.

"Join Our Mission to End Period Poverty | Always®." Join Our Mission to End Period Poverty | Always®. https://always.com/en-us/about-us/end-period-poverty.

Kampala, Dorah Egunyu in. "A Bleeding Shame: Why Is Menstruation Still Holding Girls Back?" The Guardian. May 28, 2014. https://www.theguardian.com/global-development-professionals-network/2014/may/28/menstruation-girls-education-uganda-sanitation.

Kane, Jessica. "This Is The Price Of Your Period." HuffPost. December 07, 2017. https://www.huffpost.com/entry/period-cost-lifetime_n_7258780.

Scranton, Alexandra. "Chem Fatale." Women's Voices for the Earth. November 2013. https://www.womensvoices.org/wp-content/uploads/2013/11/Chem-Fatale-Report.pdf.

Scranton, Alexandra. "Chem Fatale ." Women's Voices for the Earth: Creating a Toxic-Free Future. Women's Voices for the Earth, November 2013. https://www.womensvoices.org/wp-content/uploads/2013/11/Chem-Fatale-Report.pdf.

Thorpe, JR. "It's Actually Possible To Use Pads & Tampons Sustainably - Here's How." Bustle. April 25, 2018. Accessed June 10, 2019. https://www.bustle.com/p/tampon-disposal-other-period-habits-impact-the-environment-in-some-scary-ways-8823338.

"What We Do." Disruptor Awards. https://www.disruptorawards.com/about.

CHAPTER 11

"15 Ways to Stop Microfiber Pollution Now." Plastic Pollution Coalition. March 02, 2017. https://www.plasticpollutioncoalition.org/pft/2017/3/2/15-ways-to-stop-microfiber-pollution-now.

Amed, Imran, Achim Berg, Leonie Brantberg, and Saskia Hedrich. "The State of Fashion 2017." McKinsey & Company. December 2016. https://www.mckinsey.com/industries/retail/our-insights/the-state-of-fashion.

"Are Your Clothes Poisoning You?" The Peahen. February 5, 2015. http://thepeahen.com/are-your-clothes-poisoning-you/.

"Complimentary White Paper." QIMA. https://www.qima.com/whitepaper/quick-guide-azo-dyes.

"Fact Sheet: Nonylphenols and Nonylphenol Ethoxylates." EPA. November 02, 2016. https://www.epa.gov/assessing-and-managing-chemicals-under-tsca/fact-sheet-nonylphenols-and-nonylphenol-ethoxylates#risks.

Conca, James. "Making Climate Change Fashionable - The Garment Industry Takes On Global Warming." Forbes. December 03, 2015. https://www.forbes.com/sites/jamesconca/2015/12/03/making-climate-change-fashionable-the-garment-industry-takes-on-global-warming/#128e206a79e4.

"Fast Fashion: Definition of Fast Fashion by Lexico." Lexico Dictionaries | English. Lexico Dictionaries, n.d. https://www.lexico.com/en/definition/fast_fashion.

"Get the Facts: NPEs (Nonylphenol Ethoxylates)." Safer Chemicals, Healthy Families. Safer Chemicals, Healthy Families, n.d. https://saferchemicals.org/chemicals/npes-nonylphenol-ethoxylates/.

Haque, Moinul. "Poor Wages Force Bangladesh RMG Workers to Skip Meals." New Age | The Most Popular Outspoken English Daily in Bangladesh. February 27, 2019. http://www.newagebd.net/article/65906/poor-wages-force-rmg-workers-to-skip-meals.

Moss, Rebecca. "Mass Fainting and Clothing Chemicals." How We Get To Next. September 29, 2016. https://howwegettonext.com/mass-fainting-and-clothing-chemicals-c352b94bb54b.

"Phthalates and DEHP." Healthcare Without Harm. Healthcare Without Harm, n.d. https://noharm-uscanada.org/issues/us-canada/phthalates-and-dehp.

"Polyester." How Products Are Made. http://www.madehow.com/Volume-2/Polyester.html.

Shelbizleee. YouTube. YouTube, April 24, 2019. https://www.youtube.com/watch?v=ukHD64Ci9dQ&frags=pl,wn.

Thompson, Clay. "Ask Clay: Does Arizona Cotton Really Use That Much Water?" Azcentral. June 27, 2016. https://www.azcentral.com/story/opinion/op-ed/claythompson/2016/06/27/ask-clay-cotton-water-hog/86449070/.

"Toxic Threads: The Big Fashion Stitch-Up." Greenpeace. https://www.greenpeace.org/archive-sweden/Global/sweden/miljogifter/dokument/2012/Toxic_Threads_The_Big_Fashion_Stitch_Up.pdf.

"What Material Is the Guppyfriend Made of and Is It Recyclable?" GUPPYFRIEND. http://guppyfriend.com/en/.

CHAPTER 12

Asprey, Dave. *Head Strong: The Bulletproof Plan to Activate Untapped Brain Energy to Work Smarter and Think Faster-in Just Two Weeks.* Harper Wave, 2017.

Brown, Richard, Gaétan Chevalier, and Michael Hill. "Grounding after Moderate Eccentric Contractions Reduces Muscle Damage." Open Access Journal of Sports Medicine. September 21, 2015. https://www.ncbi.nlm.nih.gov/pmc/articles/PMC4590684/.

"Creatine Kinase (Blood)." Creatine Kinase (Blood) - Health Encyclopedia - University of Rochester Medical Center. https://www.urmc.rochester.edu/encyclopedia/content.aspx?ContentTypeID=167&ContentID=creatine_kinase_blood.

"Does Grounding Really Work?" Bulletproof. December 12, 2017. https://blog.bulletproof.com/does-grounding-work/.

Ghaly, Maurice, and Dale Teplitz. "The Biologic Effects of Grounding the Human Body during Sleep as Measured by Cortisol Levels and Subjective Reporting of Sleep, Pain, and Stress." Journal of Alternative and Complementary Medicine (New York, N.Y.). October 2004. https://www.ncbi.nlm.nih.gov/pubmed/15650465.

Murphy, Sam. "Why Barefoot Is Best for Children." The Guardian. August 09, 2010. https://www.theguardian.com/lifeandstyle/2010/aug/09/barefoot-best-for-children.

Wells, Katie. "Earthing & Grounding: Legit or Hype? (How to & When Not To)." Wellness Mama. January 23, 2019. https://wellnessmama.com/5600/earthing-grounding/.

"What Is EZ Water and Why Do I Have to Get Naked In the Sun to Make It?" Bulletproof. February 18, 2019. https://blog.bulletproof.com/ez-water/.

CHAPTER 13

Barclay, Eliza. "Your Grandparents Spent More Of Their Money On Food Than You Do." NPR. March 02, 2015. https://www.npr.org/sections/thesalt/2015/03/02/389578089/your-grandparents-spent-more-of-their-money-on-food-than-you-do).

Batts, Vicki. "Farmers at Higher Risk of Developing Various Cancers Caused by Pesticide Exposure." NaturalNews. August 24, 2016. https://www.naturalnews.com/055074_pesticides_cancer_toxic_chemicals.html.

Buettner, Dan. In *The Blue Zones Solution: Eating and Living Like the World's Healthiest People*, 604–5. Washington D.C.: National Geographic Society, 2015.

Charles, Dan. "Are Organic Vegetables More Nutritious After All?" NPR. July 11, 2014. https://www.npr.org/sections/thesalt/2014/07/11/330760923/are-organic-vegetables-more-nutritious-after-all.

Chowder, Duke's Seafood &. "Environmental Impact of Salmon Decline: This Isn't Just about Fish | Provided by Duke's Seafood & Chowder." The Seattle Times. February 07, 2018. https://www.seattletimes.com/sponsored/environmental-impact-of-salmon-decline-this-isnt-just-about-fish/.

"Cost of Organic Food - Consumer Reports." Cost of Organic Food - Consumer Reports. https://www.consumerreports.org/cro/news/2015/03/cost-of-organic-food/index.htm.

"Daily News 27 / 04 / 2018." European Commission - PRESS RELEASES - Press Release - Daily News 27 / 04 / 2018. http://europa.eu/rapid/press-release_MEX-18-3583_en.htm.

Environmental Working Group. "EWG's 2019 Shopper's Guide to Pesticides in Produce™."
EWG's 2019 Shopper's Guide to Pesticides in Produce | Summary. https://www.ewg.org/
foodnews/summary.php.

Fraser, Carly. "France Becomes The First Country to Ban All Five Pesticides
Linked to Bee Deaths." Live Love Fruit. May 13, 2019. https://livelovefruit.com/
france-bans-neonicotinoid-pesticides/.

Gallup, Inc. "Forty-Five Percent of Americans Seek Out Organic Foods." Gallup.com.
https://news.gallup.com/poll/174524/forty-five-percent-americans-seek-organic-foods.aspx.

"GLOBAL ORGANIC AREA REACHES ANOTHER ALL-TIME HIGH." IFOAM Organ-
ics International. IFOAM Organics International, February 13, 2019. https://www.ifoam.
bio/en/news/2019/02/13/world-organic-agriculture-2019.

Grossman, Elizabeth. "Declining Bee Populations Pose a Threat to Global
Agriculture." Yale E360. April 30, 2013. https://e360.yale.edu/features/
declining_bee_populations_pose_a_threat_to_global_agriculture.

Hotakainen, Rob, and E&E NewsJan. "Common Pesticides Threaten Salmon, U.S.
Fisheries Agency Concludes." Science. January 12, 2018. https://www.sciencemag.org/
news/2018/01/common-pesticides-threaten-salmon-us-fisheries-agency-concludes.

"List of US States By Size." List of US States By Size, In Square Miles, n.d. https://state.1key-
data.com/states-by-size.php.

"News & Findings." Agricultural Health Study. https://aghealth.nih.gov/news/.

"Only 1 Percent of US Farmland Is Certified Organic. Why Aren't More Farmers Making
the Switch?" Genetic Literacy Project. January 05, 2019. https://geneticliteracyproject.
org/2017/03/02/1-percent-us-farmland-certified-organic-arent-american-farmers-making-
switch/.

"Organic Market Overview." USDA ERS - Organic Market Overview. https://www.ers.usda.gov/
topics/natural-resources-environment/organic-agriculture/organic-market-overview.aspx.

Pearson, Gwen. "Bees Are Great at Pollinating Flowers-But So Are Vibrators." Wired.
June 06, 2017. https://www.wired.com/2015/05/bees-great-pollinating-flowers-vibrators/.

Roseboro, Ken. "Why Is Glyphosate Sprayed on Crops Right Before Harvest?" EcoWatch.
January 31, 2019. https://www.ecowatch.com/roundup-cancer-1882187755.html.

"Schedule for Review of Neonicotinoid Pesticides." EPA. June 06, 2019. https://www.epa.
gov/pollinator-protection/schedule-review-neonicotinoid-pesticides.

Sidder, Aaron. "New Map Highlights Bee Population Declines Across the U.S."
Smithsonian.com. February 23, 2017. https://www.smithsonianmag.com/smart-news/
new-map-highlights-bee-population-declines-across-us-180962268.

"Summary from the Health Advisory (HA) for Dacthal and Dacthal Degradates (Tetra-chloroterephthalic Acid and Monomethyl Tetrachloroterephthalic Acid) ." EPA. EPA, n.d. https://www.epa.gov/sites/production/files/2014-09/documents/summary_from_the_ha_for_dacthal_and_dacthal_degradates_tpa_and_mtp.pdf#targetText=There are no heath data,is less toxic than dacthal.

CHAPTER 14

Ali, Fareeha. "US Ecommerce Sales Grow 15.0% in 2018." Digital Commerce 360, February 28, 2019. https://www.digitalcommerce360.com/article/us-ecommerce-sales/.

Bird, Jon. "What A Waste: Online Retail's Big Packaging Problem." Forbes. Forbes Magazine, July 29, 2018. https://www.forbes.com/sites/jonbird1/2018/07/29/what-a-waste-online-retails-big-packaging-problem/#581d91d7371d.

DePillis, Lydia. "America's Addiction to Absurdly Fast Shipping Has a Hidden Cost." CNN. Cable News Network, July 15, 2019. https://www.cnn.com/2019/07/15/business/fast-shipping-environmental-impact/index.html.

"Frustration-Free Packaging." US About Amazon, May 8, 2018. https://www.aboutamazon.com/sustainability/packaging/frustration-free-packaging.

Godlewski, Nina. "Amazon Employees Have Resorted to Urinating in Trash Cans in Some Warehouses." Newsweek. Newsweek, February 26, 2019. https://www.newsweek.com/amazon-drivers-warehouse-conditions-workers-complains-jeff-bezos-bernie-1118849.

Laurenthomas. "Watch out, Retailers. This Is Just How Big Amazon Is Becoming." CNBC. CNBC, July 13, 2018. https://www.cnbc.com/2018/07/12/amazon-to-take-almost-50-per-cent-of-us-e-commerce-market-by-years-end.html.

"United States: E-Commerce Share of Retail Sales 2021." Statista. Statista, n.d. https://www.statista.com/statistics/379112/e-commerce-share-of-retail-sales-in-us/.

CHAPTER 15

Andrews, David. "Teflon-Killing Canaries and the American Dream." EWG. EWG, May 1, 2015. https://www.ewg.org/enviroblog/2015/05/teflon-killing-canaries-and-american-dream.

Burke, Caroline. "What Are PFAS? Your Takeout Containers Could Have Them & You Wouldn't Know." Bustle. Bustle, August 20, 2019. https://www.bustle.com/p/what-are-pfas-your-takeout-containers-could-have-them-you-wouldnt-know-18684267.

"Buyer Guide: Steel PDF." Environmental Defense. Environmental Defense, n.d.

Chrobak, Ula. "Eco-Friendly Packaging Could Be Poisoning Our Com-post." Popular Science. Popular Science, May 30, 2019. https://www.popsci.com/compostable-packaging-PFAS/.

"Detoxing From Heavy Metals with Wendy Myers." The Energy Blueprint. The Energy Blueprint, August 12, 2019. https://www.theenergyblueprint.com/detoxing-from-heavy-metals-wendy-myers/.

"Does Cooking with Cast Iron Pots and Pans Add Iron to Our Food?" Go Ask Alice. Go Ask Alice, n.d. https://goaskalice.columbia.edu/answered-questions/does-cooking-cast-iron-pots-and-pans-add-iron-our-food.

Grinvalsky, Jim. "Molded Fiber Food Packaging with PFAS: Is It Safe and Compostable?" EBP Supply. EBP Supply, August 16, 2019. https://www.ebpsupply.com/blog/molded-fiber-food-packaging-safe-compostable-contains-pfas.

"Healthier Food Serviceware Choices." Center for Environmental Health. CEH, May 2019. https://www.ceh.org/wp-content/uploads/PFAS-in-Foodware-Infographic.pdf.

"How Teflon™ Is Made." Chemours. Chemours, n.d. https://www.chemours.com/Teflon/en_US/products/safety/how_its_made.html.

Kelly, Sharon. "DuPont's Deadly Deceit: The Decades-Long Cover-up behind the 'World's Most Slippery Material.'" Salon. Salon.com, January 5, 2016. https://www.salon.com/2016/01/04/teflons_toxic_legacy_partner/.

Measom, Cynthia. "What Are the Dangers of Copper Cookware?" Hunker. Hunker, n.d. https://www.hunker.com/12003785/what-are-the-dangers-of-copper-cookware.

OK International. "Cookware made with scrap metal contaminates food: Study across 10 countries warns of lead and other toxic metals." ScienceDaily. www.sciencedaily.com/releases/2017/01/170123110345.htm.

Perez, Sandrine. "Is Stainless Steel Cookware Safe?" Nourishing Our Children. Nourishing Our Children, September 28, 2017. https://nourishingourchildren.org/2017/09/25/stainless-steel-cookware/.

"PFAS Contamination of Water." State of Rhode Island: Department of Health. State of Rhode Island, n.d. http://www.health.ri.gov/water/about/pfas/.

"PFAS Master List of PFAS Substances." EPA. Environmental Protection Agency, n.d. https://comptox.epa.gov/dashboard/chemical_lists/pfasmaster.

Sharp, Renee, and J. Paul Pestano. "Water Treatment Contaminants:" EWG. EWG, February 27, 2013. https://www.ewg.org/research/water-treatment-contaminants.

Spencer, Keith A. "The Chemical Industry Doesn't Want You to Be Afraid of Teflon Pans. You Should Be." Salon. Salon.com, February 4, 2018. https://www.salon.com/2018/02/04/the-chemical-industry-doesnt-want-you-to-be-afraid-of-teflon-pans-you-should-be/.

Tchounwou, Paul B, Clement G Yedjou, Anita K Patlolla, and Dwayne J Sutton. "Heavy Metal Toxicity and the Environment." PMC. U.S. National Library of Medicine, 2012. https://www.ncbi.nlm.nih.gov/pmc/articles/PMC4144270/.

"Teflon's Toxic Legacy: DuPont Knew for Decades It Was Contaminating Water Supplies." EcoWatch. EcoWatch, January 4, 2016. https://www.ecowatch.com/teflons-toxic-legacy-dupont-knew-for-decades-it-was-contaminating-wate-1882142514.html.

"Why Ceramic Cookware Is Safer Than Stainless Steel." Xtrema Pure Ceramic Cookware. Xtrema.com, July 26, 2018. https://www.xtrema.com/blogs/blog/why-ceramic-cookware-is-safer-than-stainless-steel.

CHAPTER 16

"1,4-DIOXANE." EWG Skin Deep Cosmetics Database. EWG, n.d. http://www.ewg.org/skindeep/ingredient/726331/1,4-DIOXANE/#.

Bhattacharyya, Sumit, Leo Feferman, and Joanne K. Tobacman. "Increased Expression of Colonic Wnt9A through Sp1-Mediated Transcriptional Effects Involving Arylsulfatase B, Chondroitin 4-Sulfate, and Galectin-3." Journal of Biological Chemistry. Journal of Biological Chemistry, June 20, 2014. http://www.jbc.org/content/289/25/17564.abstract.

Boronow, Katherine E., Julia Green Brody, Laurel A. Schaider, Graham F. Peaslee, Laurie Havas, and Barbara A. Cohn. "Serum Concentrations of PFASs and Exposure-Related Behaviors in African American and Non-Hispanic White Women." Nature News. Nature Publishing Group, January 8, 2019. https://www.nature.com/articles/s41370-018-0109-y.

Cassidy, Emily. "Did You Know That Monsanto's Glyphosate Doubles the Risk of Cancer?" EWG. EWG, October 6, 2015. https://www.ewg.org/agmag/2015/10/monsanto-s-gmo-herbicide-doubles-cancer-risk.

"CFR - Code of Federal Regulations Title 21." FDA. FDA, n.d. https://www.accessdata.fda.gov/scripts/cdrh/cfdocs/cfcfr/cfrsearch.cfm?fr=355.50.

Ciancio, Sebastian G. "Baking Soda Dentifrices and Oral Health." The Journal of the American Dental Association. JADA, November 2017. https://jada.ada.org/article/S0002-8177(17)30822-X/fulltext.

Cornucopia. The Cornucopia Institute, 2016. https://www.cornucopia.org/wp-content/uploads/2016/08/toothpaste-report-web.pdf.

Darbre, Philippa D, and Philip W Harvey. "Paraben Esters: Review of Recent Studies of Endocrine Toxicity, Absorption, Esterase and Human Exposure, and Discussion of Potential Human Health Risks." Journal of applied toxicology : JAT. U.S. National Library of Medicine, July 2008. https://www.ncbi.nlm.nih.gov/pubmed/18484575.

"FDA Issues Final Rule on Safety and Effectiveness of Antibacterial Soaps." U.S. Food and Drug Administration. FDA, September 2, 2016. https://www.fda.gov/news-events/press-announcements/fda-issues-final-rule-safety-and-effectiveness-antibacterial-soaps.

"FORMALDEHYDE." EWG Skin Deep Cosmetics Database. EWG, n.d. http://www.ewg. org/skindeep/ingredient/702500/FORMALDEHYDE/#.

Hari, Vani. "Is Your Toothpaste Full Of Carcinogens? Check This List..." Food Babe. Food Babe, October 5, 2016. https://foodbabe.com/ brushing-teeth-carcinogens-dirty-truth-toothpaste/.

Healy, Melissa. "Triclosan Could Be Really Harmful to Your Gut, and It's Probably in Your Toothpaste." Los Angeles Times. Los Angeles Times, May 31, 2018. https://www. latimes.com/science/sciencenow/la-sci-sn-triclosan-cancer-risk-20180531-story.html.

Herlofson, BB, and P Barkvoll. "Sodium Lauryl Sulfate and Recurrent Aphthous Ulcers. A Preliminary Study." Acta odontologica Scandinavica. U.S. National Library of Medicine, October 1994. https://www.ncbi.nlm.nih.gov/pubmed/7825393.

"Impact of Fluoride on Neurological Development in Children." Harvard T.H. Chan. Harvard , July 25, 2012. https://www.hsph.harvard.edu/news/features/ fluoride-childrens-health-grandjean-choi/.

Limitone, Julia. "Vegan Toothpaste Pill Aims to Cut Plastics in Landfills." Fox Business. Fox Business, February 18, 2019. https://www.foxbusiness.com/small-business/ this-pill-is-really-vegan-toothpaste-that-keeps-plastic-from-landfills.

Narang, Neha, and Jyoti Sharma. "SUBLINGUAL MUCOSA AS A ROUTE FOR SYSTEMIC DRUG DELIVERY." Innovare Academic Sciences, 2011. https://innovareacademics.in/journal/ijpps/Vol3Suppl2/1092.pdf.

"NITROSAMINES." EWG's Skin Deep Cosmetics Database. EWG, n.d. https://www.ewg. org/skindeep/ingredient/726336/NITROSAMINES/.

Ogimoto, Mami, Yoko Uematsu, Kumi Suzuki, Junichiro Kabashima, and Mitsuo Nakazato. "Survey of Toxic Heavy Metals and Arsenic in Existing Food Additives (Natural Colors)." Shokuhin eiseigaku zasshi. Journal of the Food Hygienic Society of Japan. U.S. National Library of Medicine, October 2009. https://www.ncbi.nlm.nih.gov/pubmed/19897953.

"Palm Oil." WWF. World Wildlife Fund, n.d. https://www.worldwildlife.org/ industries/palm-oil?fbclid=IwAR2ypLRR93PowWGPsaWBNv6JGZOT2XsMm4GRl-utS4t67J9o-VVANLKG12U.

"Summary of Studies on Food Dyes." Center for Science in the Public Interest. Center for Science in the Public Interest , n.d. https://cspinet.org/sites/default/files/attachment/ dyes-problem-table.pdf.

CHAPTER 17

"Adipose Tissue." ScienceDaily. https://www.sciencedaily.com/terms/adipose_tissue.htm.

Cecchini, Marie A., David E. Root, Jeremie R. Rachunow, and Phyllis M. Gelb. "Health Status of Rescue Workers Improved by Sauna Detoxification." Arthritis Trust. Arthritis Trust of America, April 2006. http://arthritistrust.info/wp-content/uploads/2013/03/Chemical-Exposure-at-World-Trade-Center.pdf.

Gadacz, René R. "Sweat Lodge." *The Canadian Encyclopedia*, The Canadian Encyclopedia, 7 Feb. 2006, https://www.thecanadianencyclopedia.ca/en/article/sweat-lodge.

Lee, Y.-M., K.-S. Kim, D. R. Jacobs, and D.-H. Lee. "Persistent Organic Pollutants in Adipose Tissue Should Be Considered in Obesity Research." Obesity Reviews. December 02, 2016. https://onlinelibrary.wiley.com/doi/full/10.1111/obr.12481.

Rose, Sahara. "What's a Panchakarma? Panchakarma Ayurvedic Detoxification Treatments Explained." Eat Feel Fresh. December 15, 2016. https://eatfeelfresh.com/whats-panchakarma-panchakarma-ayurvedic-detoxification-treatments-explained/.

"Toxic Chemicals Released by Industries This Year, Tons." Worldometers. https://www.worldometers.info/view/toxchem/.

CHAPTER 18

Bosworth, Mark. "Why Finland Loves Saunas." BBC News. October 01, 2013. https://www.bbc.com/news/magazine-24328773.

"EMF Radiation? Should You Really Be Concerned?" Jill Carnahan, MD. March 22, 2018. https://www.jillcarnahan.com/2017/10/29/emf-radiation/.

Julia, Camille. "The Ultimate Helsinki Sauna Crawl." Becoming Fully Human. March 01, 2019. https://www.becomingfullyhuman.ca/the-bfh-guide-to/2019/2/14/the-ultimate-helsinki-sauna-crawl?rq=sauna.

Laukkanen, Tanjaniina, Hassan Khan, and Francesco Zaccardi. "Association Between Sauna Bathing and Fatal Cardiovascular and All-Cause Mortality Events." JAMA Internal Medicine. American Medical Association, April 1, 2015. https://jamanetwork.com/journals/jamainternalmedicine/fullarticle/2130724.

Nelson, Marilee. "Sunning: Take Advantage of the Summer Sun to Off-gas VOCs." Branch Basics, May 18, 2018. https://branchbasics.com/blog/sunning-how-to-take-advantage-of-the-summer-sun-by-outgassing/.

"What Is Photobiomodulation? Photobiomodulation Process." Vielight Inc. https://vielight.com/photobiomodulation/.

CHAPTER 19

"America's $165 Billion Food-waste Problem." CNBC. July 17, 2015. https://www.cnbc.com/2015/04/22/americas-165-billion-food-waste-problem.html.

Greenfield, Rob. "How To End The Food Waste Fiasco | Rob Greenfield | TEDx-Teen." YouTube. TedX Talks, February 2, 2016. https://www.youtube.com/watch?v=w96osGZaS74&frags=pl,w.

"Legislative Work." Zero Waste Washington, n.d. https://zerowastewashington.org/legislative-work/.

Mosbergen, Dominique. "China No Longer Wants Your Trash. Here's Why That's Potentially Disastrous." HuffPost. HuffPost, January 25, 2018. https://www.huffpost.com/entry/china-recycling-waste-ban_n_5a684285e4b0dc592a0dd7b9.

"United Airlines Makes History with Launch of Regularly Scheduled Flights Using Sustainable Biofuel." United. United Airlines, Inc., January 31, 2019. https://hub.united.com/united-launch-flights-sustainable-biofuel-2567373100.html.

INDEX

—

A

Alzheimer's 29, 203
amalgam 28-29
Amazon 91, 185-187, 190, 192-193
Amour Vert 17, 147-154, 264
American Sustainability Business Counsel (ASBC) 50, 59-60
Asprey, Dave 160-161
avobenzone 18
Ayurveda 74, 222

B

Baird, Katherine 55-58
Balti, Linda 17, 147-154, 264
bees 42, 173-176, 180

M

makeup 7, 12, 16, 21, 49, 83, 85-87, 263

Marcus, Pam 16, 96-100

McCormick, Lindsay 17, 215-220

mercury 28-29, 42, 200, 223

metal 19, 35, 80-81, 101, 199-206, 210, 223, 231, 239-240, 263

microfiber 144, 155, 156

microplastic 95, 263

mitochondria 160, 200, 238

modal 148

Monsanto 51-52

N

naturopath 72, 74, 116, 206

naturopathic 73-74, 77, 106, 108, 225

neonicotinoids 174-176

nitrosamines 119-120, 128, 210

non-stick 195-198, 205-206

Nonylphenol Ethoxylates (NPEs) 145

O

organic 8, 11, 17, 20-21, 51, 58, 62-64, 72, 75-77, 79, 81, 119, 127, 141, 143, 147, 151, 155, 167-173, 176-177, 180-183, 188, 192, 223, 240, 253, 258

Orme, Shelbi 144, 247, 258, 259

P

palm oil 217

Patagonia 189, 265-266

Post-Birth Control Syndrome (PBCS) 111

PCC 17, 94, 176-182

Perzow, Barry 16, 71-80

pesticide 51, 56, 75, 83, 120, 131, 143, 145, 151, 170-176, 180-181, 223, 225

petrochemical 122-122, 239

PFAS 55, 195-199, 207, 220

PFOA 197-199

Pharmaca 16, 71, 72, 76-80, 128

photobiomodulation 236-238

the pill 106-111, 113, 115, 118, 214

Pineault, Nick 36-37, 39-40, 45

plastic 7, 20, 91-97, 99-101, 138, 155-156, 163, 180, 183, 186, 192, 214-216, 218-220, 248, 250-251, 254, 260

polyester 144-145, 155-156

precautionary principle 27, 42

R

radiation 36-45, 48

radiofrequency 36-38

Rechberg, Natalie 17, 113-118

recycling 96, 249-251, 254

Richards, Brian 18, 236-241

Made in the USA
Columbia, SC
03 January 2020

86177221R00165